THE PHYSICS OF
SUPERHEROES

THE PHYSICS OF
SUPERHEROES

James Kakalios

GOTHAM BOOKS

GOTHAM BOOKS
Published by Penguin Group (USA) Inc.
375 Hudson Street, New York, New York 10014, U.S.A.
Penguin Group (Canada), 90 Eglinton Avenue East, Toronto, Ontario, Canada
M4P 2Y3 (a division of Pearson Penguin Canada Inc.); Penguin Books Ltd, 80
Strand, London WC2R 0RL, England; Penguin Ireland, 25 St Stephen's Green,
Dublin 2, Ireland (a division of Penguin Books Ltd); Penguin Group (Australia),
250 Camberwell Road, Camberwell, Victoria 3124, Australia (a division of
Pearson Australia Group Pty Ltd); Penguin Books India Pvt Ltd, 11
Community Centre, Panchsheel Park, New Delhi–110 017, India; Penguin
Group (NZ), cnr Airborne and Rosedale Roads, Albany, Auckland 1310, New
Zealand (a division of Pearson New Zealand Ltd); Penguin Books (South
Africa) (Pty) Ltd, 24 Sturdee Avenue, Rosebank, Johannesburg 2196, South
Africa

Penguin Books Ltd, Registered Offices: 80 Strand, London WC2R 0RL,
England

Published by Gotham Books, a division of Penguin Group (USA) Inc.

First printing, October 2005
10 9 8 7 6 5 4 3

LIBRARY OF CONGRESS CATALOGING-IN-PUBLICATION DATA HAS BEEN APPLIED
FOR.

ISBN 1-592-40146-5

Printed in the United States of America
Set in Trump Mediaeval and Comic
Designed by Sabrina Bowers

TO THERESE

CONTENTS

SECTION 2–ENERGY–HEAT AND LIGHT

SECTION 3—MODERN PHYSICS

SECTION 4—WHAT HAVE WE LEARNED?

FOREWORD

WHILE WILE E. COYOTE is by no stretch of the imagination a superhero, I have to admit that it was this hapless villain—who escaped almost certain death episode after episode while continuing to fruitlessly chase the Road Runner with almost Sisyphean intensity day in and day out—who first got me thinking about the physics of illustrated characters. Even as a relatively young boy hooked on television, I suspected there was something fishy whenever I saw Wile E. run off a cliff and hover indefinitely until the moment he realized there was no solid ground underneath. Somehow it seemed to me even then that gravity should continue to work, whether or not one was conscious of it.

I bring this example up, in spite of the fact that it involves no superheroes, and in fact involves a television cartoon rather than a comic book figure, because it illustrates a point that has become central to the way I think about teaching physics: Few things are more memorable than confronting one's own misconceptions. Indeed, some among those of us who study "physics education" for a living suggest that it is only by directly encouraging students to run up against their own misconceptions that one can hope to truly cause them to internalize what one is teaching them. I don't know whether this is true or not, but I do know that if you want to reach out to understand popular misconceptions, then *exploiting* where we get our cultural perspectives from is a good place to start. And if that means borrowing from *Superman*, or *Star Trek*, I am all for it!

Now, I don't want you to think that I bring up comic books and popular misconceptions in the same paragraph because I want to denigrate the former. Far from it! Indeed, the comics sometimes actually get it right, and as James Kakalios describes in his introduction to this far-reaching journey from the gravity of Krypton to the quantum mechanics of the X-Men, students often seem to grumble about how the standard examples from his introductory physics class have nothing to do with the real world they will encounter upon graduation. But when they are instead introduced to the physics of superheroes, this complaint never arises!

One might initially wonder whether Superman might seem more real to students than pulleys, ropes, and inclined planes. But the real reason students don't complain is undoubtedly that the comic book examples are *fun*, while inclined planes aren't. And that is perhaps one of the most useful reasons for thinking about the physics of superheroes. Not only can you imagine, and be introduced to, lots of interesting physics, from everyday phenomena to esoteric modern subjects, but it is actually fun to think about. Moreover, while subjects like Quantum Mechanics might seem intimidating, who could be intimidated by cute Kitty Pryde?

Some who remember the comic books that enthralled them as young people might also recall a sense of wistfulness in pondering whether our own world could ever capture the excitement and drama of the worlds of comic book superheroes. In truth, however, it is far *more* interesting and exciting, if only we open up our minds to the hidden wonders of nature that science has revealed to us over the past four hundred or so years. Truth is far stranger than fiction, even comic book fiction. And finding out why is all part of the fun.

Lawrence M. Krauss
Cleveland, Ohio, April 2005

PREFACE

I WAS A COMIC-BOOK FAN as a kid, but like many of those who have come before and after me, I abandoned the hobby in high school upon discovering girls. My mother, following the standard script, used this opportunity to throw my collection away. I renewed my comic-book-reading habit years later, in graduate school, as a way to relieve the stress of working on my dissertation. Now as an adult I've rebuilt much of my comic-book collection (or, as my wife refers to it, "the fire hazard"), but to be safe, my mother is not allowed near it.

Back in 1998 the University of Minnesota, where I am a physics professor, introduced a new type of class termed "freshman seminars." These are small, seminar-type classes open to entering students, and while they are for college credit, they are not tied to any particular curriculum. Professors are encouraged to develop classes on unconventional topics, and freshman seminars on Bio-Ethics and the Human Genome; The Color Red (a chemistry class); Trade and the Global Economy; and Complex Systems: From Sandpiles to Wall Street, are among the many classes offered. In 2001 I introduced a class originally entitled Everything I Know About Science I Learned from Reading Comic Books. This is an actual physics class, treating most of the topics traditionally covered, but rather than employing illustrations of masses on springs, or blocks sliding down inclined planes, all of the examples came from the four-color adventures of costumed superheroes, and focused in particular on those situations where the comic books got their physics right.

The present book, while inspired by this class, is not a textbook per se. It is written for the nonspecialist who is interested in a relatively pain-free way to learn about the basic physics concepts underlying our modern technological lifestyle. Topics such as forces and motion, conservation of energy, thermodynamics, electricity and magnetism, quantum mechanics, and solid-state physics are discussed, and real-world applications such as automobile airbags, transistors, and microwave ovens are explained. I hope you will be so busy enjoying this superhero ice cream sundae that you won't realize that I am sneakily getting you to eat your spinach at the same time.

This book is intended for both longtime comic-book fans and those who can't tell Batman from Man-Bat. I have therefore described the history and background of the comic-book heroes discussed here. In order to describe the physics connected with certain superheroes or story lines, I have had to summarize key plot points in various comic books. Therefore, for those who have not yet read these classics, the following two words apply to this entire book: "Spoiler Alert."

Readers interested in consulting the source materials considered here will find citations to the comic books discussed in the text at the end of the book. I have listed the original comic-book information and, whenever possible, where the issue can be found in a recent reprint volume at bookstores and comic shops. The dates listed for a given comic book, printed on its cover, are not when the issue first appeared on the newsstands. To extend shelf life, the date listed indicated when the comic was to be returned to the publisher for credit, and not when it became available for purchase. In an effort to attract new readers who desire first issues as collectors' items, comic books will occasionally restart their numbering scheme while keeping the name of the comic unchanged. If not otherwise noted, issue numbers refer to the first volume of a comic. I have listed, where known, the writer and artist for each comic listed in the endnotes. My omission of the inkers should not be construed as denigrating their contribution to the finished comic (I most certainly do not believe that such a job is equivalent to "tracing"), but rather a reflection of the fact that the artist along with the writer typically have the primary responsibility for the physics in a given comic book scene.

Any discussion of physics in comic books naturally invites the scrutiny of physicists as well as comic-book fans, both of who are known for their . . . attention to detail. Each of the incidents I selected happened to illustrate a particular physics principle. Sometimes the very next issue would contain a scene contradicting the physically plausible manifestation of a superpower described here. When considering characters that have starred in multiple comic books for over half a century, it is a sure bet that there will be counter-examples for any statement I make. Consequently, while examining the physics associated with a superhero's powers will, in many cases, provide a better appreciation for their talents, my comrades in fandom are advised that this book is not intended to provide definitive accounts of any character's power or adventures.

Similarly, my physics colleagues are warned that this book is for a non-expert audience. I have attempted to keep things simple, while acknowledging the rough edges and complications of the real world. A complete discussion of many of the topics considered here could easily expand to fill several volumes and would provide a concrete illustration of Dr. Manhattan's final words at the conclusion of Alan Moore's and Dave Gibbons' *Watchmen*: "Nothing ends, Adrian. Nothing ever ends."

The language that describes the physical world is mathematical in nature. Why this should be so is a deep, philosophical question (the physicist Eugene Wigner refered to the "unreasonable effectiveness of mathematics" in accounting for nature's properties) that has puzzled and stirred all those who have studied it. It is tempting, in a book involving comic-book superheroes, to avoid even the slightest whiff of mathematics. However, that would be a cheat, worse than omitting any reproductions of artwork in a book about Picasso, or not providing a CD of recordings for a book about the History of Jazz, because mathematics is necessary in any thorough discussion of physics.

The reader may protest that they do not understand math, or cannot think mathematically. But for this book, all that is required is that one recognizes that $\frac{1}{2}+\frac{1}{2}=1$. That's it, that two halves equal a whole. If you are comfortable with $\frac{1}{2}+\frac{1}{2}=1$, then writing it as $2\times(\frac{1}{2})=\frac{2}{2}=1$ (that is, two multiplied by one half) should cause no concern, because obviously two halves equal one. It seems so simple you may be surprised to discover that we have already been doing algebra (and you thought you'd never use that again after high school!).

As many students have long suspected, there is a trick to algebra, and the trick is the following: If one has an equation describing a true statement, such as $1=1$, then one can add, subtract, multiply, or divide (excepting division by zero) the equation by any number we wish, and as long as we do it to both the left- and right-hand sides of the equation, the correctness of the equation is unchanged. So if we add 2 to both sides of $1=1$, we obtain $1+2=1+2$ or $3=3$, which is still a true statement. Dividing both sides of $1=1$ by 2 gives us $\frac{1}{2}=\frac{1}{2}$. Since $1=1$, then $\frac{1}{2}+\frac{1}{2}=1$, which in turn can be written as $\frac{2}{2}=1$. I'll make a deal with you, the reader: I won't use any mathematics more sophisticated than described in this paragraph, if you'll refrain from panicking when a mathematical equation appears. You can always glide right by the math, and your understanding will be no worse for it. But if you take a notion to calculate a velocity or a force for a situation other than considered here, you'll have the tools to do so. In any event, I promise you that there will be no quizzes at the end of the book!

THE PHYSICS OF
SUPERHEROES

INTRODUCTION

SECRET ORIGINS: HOW SCIENCE SAVED SUPERHERO COMIC BOOKS

IF I HAD EVER WONDERED if my students found studying physics to be a waste of time, all doubt was removed several years ago. I was returning from lunch to the physics building at my university when I overheard two students as they were leaving. From their expressions and the snippet of conversation I caught, it seemed that they had just had a graded exam returned to them. I'll quote here what I heard (but in the interest of decorum, I'll clean it up).

The taller student complained to his friend, "I'm going to bleeping buy low, and bleeping sell high. I don't need to know about no bleeping balls thrown off no bleeping cliffs."

There are two things we can learn from this statement: (1) the secret to financial success and (2) that the examples used in traditional physics classes strike many students as divorced from their everyday concerns.

The real world is a complicated place. In order to provide illustrations in a physics lesson that emphasize only a single concept, such as Newton's Second Law of Motion or the principle of Conservation of Energy, over the decades physics teachers have developed an arsenal of overly stylized scenarios involving projectile motion, weights on pulleys, or oscillating masses on springs. These situations seem so artificial that students inevitably lament, "When am I ever going to use this stuff in my real life?"

One trick I've hit upon in teaching physics involves using examples culled from superhero comic books that correctly illustrate various applications of physics principles. Interestingly enough, whenever I cite examples from superhero comic books in a lecture, my students *never* wonder when they will use this information in "real life." Apparently they all have plans, post-graduation, that involve Spandex and protecting the City from all threats. As a law-abiding citizen, it fills me with a great sense of security because I also know how many of my scientist colleagues could charitably be termed "mad."

* * *

I first made the connection between comic books and college education back in 1965, when for the princely sum of twelve cents I purchased *Action Comics* # 333, which featured the adventures of Superman. While not a huge fan of the Man of Steel at the time, I was seduced by the comic's cover (see fig. 1), which promised a glimpse into the inner workings of our institutions of higher learning. As a kid I was deeply curious as to what college life would be like. Now that I am a university professor, I realize that this was a premonition that once I entered college I would never get out, and that my matriculation would turn into some sort of life sentence.

A story in *Action* # 333, titled "Superman's Super Boo-Boos," featured a scene in which, as a "tribute to all he's done for humanity," Superman was to be granted an "honorary degree of Doctor of Super-Science" by Metropolis Engineering College. (I should point out that such a degree option was not offered when I enrolled in graduate school.) On the cover of this issue Superman was in a large auditorium on a college campus, and was "writing" his name on a bronze honor scroll using his heat vision. The aged faculty in attendance, wearing commencement dress, were aghast, not due to the fact that searing beams of energy were emanating from Superman's eyes but rather because the professors saw a fire-breathing dragon up on the stage instead of Superman due to an illusion cast by Superman's archnemesis, Lex Luthor. This was part of Luthor's scheme to constantly confound Superman's expectations until he was paralyzed by indecision and hence unable to stop Luthor's evil schemes.* Even as a grade-school kid I realized that this

---

* To this end Luthor publicly came to Superman's aid on several occasions, with the goal of making Superman doubt his judgment regarding Luthor's intentions. So committed was Lex Luthor to this plan that he even directly saved Superman's life when he was threatened by another crook wielding a

Fig. 1.
Cover to
Action
Comics # 333,
featuring a
scene from
Superman's
ill-fated visit
to Metropolis
Engineering
College.

depiction of college life was probably not too realistic. Neverthe-less the cover did provide two insights that, in the fullness of time, have turned out to be fairly accurate. The first is that *all* college professors, at all times, *always* wear caps and gowns. The second is that all college professors are eight-hundred-year-old white men.

Kryptonian sword capable of killing him. You might think that it would be simpler at this point to abandon his plan to "gaslight" Superman and just let the other crook kill him and be done with it, but who can truly understand the thought processes of a criminal mastermind like Lex Luthor?

While this may have been my first indication that comic books and college could coexist, it would not be my last. Over the years I have continued to enjoy reading and collecting comic books. This is not a "guilty pleasure" of mine, simply because I don't believe in "guilty" pleasures. Snobbery is just the public face of insecurity. (You like what you like, and you shouldn't feel guilty about your interests or hobbies. Unless, of course, you golf.) And in my reading I've noted that the writers and artists creating superhero comic-book stories get their science right more times than you might expect. Those not overly familiar with superhero comic books may be surprised to learn that *anything* in comic books could be scientifically correct, but one can learn a lot of science from reading comic books.

A typical example is shown in fig. 2, featuring a scene from *World's Finest* # 93 from April 1958. The big superstars at National Comics (which then became Detective Comics and is today known as DC Comics) are Superman, Batman, and Robin and each issue of *World's Finest* contained a team-up adventure of the Man of Steel and the Dynamic Duo. In this story a crook, Victor Danning, has his intelligence accidentally increased to "genius level" during a botched attempt to steal a "brain amplifier" machine. Using his enhanced mental powers, he proceeds to commit a series of "super-crimes," requiring the attention of Batman and Robin

Fig. 2. *A scene from* World's Finest # 93. *In this scene a crook whose intelligence has been artificially enhanced describes his plan to utilize the dispersion of underground shock waves in order to determine the hidden location of the Batcave.*

and Superman. After several of his schemes are narrowly foiled by our heroes, Danning decides to take a proactive approach and attempts to discover the hidden headquarters of Batman and Robin, the Batcave. (It's not really explained how this would defeat Batman and Robin, and it is presented as a given that any crook would want to learn the whereabouts of the Batcave.)

Danning instructs his henchmen to plant sticks of dynamite along the perimeter of Gotham City. Monitoring the resulting shock waves on his "radar-seismograph," Danning explains that the waves traveling through a cave will have different speeds than those moving through solid rock, and in this way the location of the Batcave can be discerned. In this example the evil genius Victor Danning is on solid scientific ground, for it is indeed true that the velocity of sound or a shock wave will depend on the density of the material through which it moves. In fact, geologists make use of this variation of the speed of sound waves in order to locate underground pockets of oil or natural gas.

The depiction in comic books of actual scientists and how they work, on the other hand, often leaves much to be desired. In the very same issue of *World's Finest*, a less realistic portrayal of scientists is presented, as shown in fig. 3. Here the inventor of the "brain amplifier," Dr. John Carr, describes his latest experiment at a scientific conference. He boasts that his device will "increase any man's

© 1958 National Periodical Publications Inc. (DC)

Fig. 3. *Another scene from* World's Finest # 93, *where "crooked ex-scientist" Victor Danning first learns of the "brain amplifier" device, along with its only drawback.*

mental power 100 times." Unfortunately, Carr points out that there's one catch, saying: "there's one ingredient that's still missing [before his machine will work], and I haven't found out yet what it is!" This is equivalent to inventing a machine that turns lead into gold or tap water into gasoline—but needs one key element (and who knows if it even exists) to work! Rarely are presentations made at physics conferences of work in such an unfinished state (at least, not intentionally). Watching this presentation is Victor Danning (this is where he was inspired to steal the brain amplifier, despite its intrinsic design flaw), who is labeled in the caption box as a "crooked ex-scientist." This part rings true, for I speak not just as a physicist but for all scientists that, once you become "crooked," we kick you out of the club and strip you of your "scientist" title.

* * *

The incorporation of scientific principles into superhero adventures is only occasionally found in stories from the 1940s (referred to by fans as the "Golden Age" of comic books) but is much more common in comics from the late 1950s and 1960s (known as the "Silver Age"). Between these two epochs lies the "Dark Age" of comic books, when sales plummeted and the very concept of superheroes came under attack by psychiatrists, educators, and congressmen. Those circumstances that led to there being two "ages" of superhero comic books are also responsible, it can be argued, for the "scientific" tone in Silver Age comics published in the post-Sputnik era. Since we will be relying on superheroes to illustrate scientific concepts for the rest of this book, it is useful to first take a moment to consider the early roots of these mystery men.

A BRIEF HISTORY OF SUPERHERO COMIC BOOKS

Before there were comic books, there were comic strips.* Weekly broadsheets in Victorian England nicknamed "penny dreadfuls" featured humorous strips in the music-hall tradition. Their popularity with poor workers was a great offense to middle-class sensi-

---

* The concept of using words and drawings to tell a story is more than five hundred years old. There are examples of woodcut "broadsheets" from the Middle Ages that use paneled borders, speedlines, and word balloons.

bilities. A fierce newspaper rivalry in the 1890s between Joseph Pulitzer and William Randolph Hearst spurred the creation of the newspaper comic strips that proved extremely popular with newly arrived immigrants who were only semi-conversant in English. These newspaper strips became highly effective weapons in the circulation wars of the time, and the innovation of printing a comic strip in color (a radical move, even if it was only the single color yellow) catapulted Hearst's papers ahead of his competitors. In fact, so closely associated would the Hearst papers become with Richard F. Outcault's visually striking and popular comic strip *The Yellow Kid* that critics would denigrate all of Hearst's sensationalistic newspapers with the sobriquet "Yellow Journalism."

Despite occasional forays into presenting comic strips in a magazine format (such as a Buster Brown comic book published in 1903, Little Nemo in 1906, and Mutt and Jeff in 1910), comic books did not become firmly established until 1933. At the time, newsstands were filled with extremely popular pulp magazines, containing an entire original novel for just ten cents. Costs were kept down, in part, by printing on the poor-quality paper that gave these journals their name. There were popular titles devoted to mystery stories, such as *Detective Fiction Weekly* and *Black Mask* (which first published Dashiell Hammett and Raymond Chandler); science fiction, such as *Amazing Stories* and *Astounding Stories* (wherein Theodore Sturgeon, Isaac Asimov, and Ray Bradbury got their start); horror and fantasy, such as *Unknown* and *Weird Tales* (the home of H. P. Lovecraft, Robert E. Howard, and even the playwright Tennessee Williams); and action/adventure titles like *The Shadow*, *The Spider*, *G-8 and His Battle Aces*, *The Mysterious Wu-Fang*, and *Doc Savage*. At the height of their popularity some pulp titles sold several hundred thousand issues per month, which even at ten cents a copy was big money during the Depression. In this highly competitive environment George Janosik, George Delacorte, Harry Wildenberg, and Maxwell C. Gaines (a schoolteacher before he became a comic-book publisher) decided to take a chance and reprint the color newspaper comic strips from the Sunday supplements onto standard tabloid-size sheets of newsprint, folding them into a 6 5/8" × 10 1/8" pamphlet (thereby establishing the standard format for comic books, which remains unchanged to this day). The comic book, *Funnies on Parade,* was distributed with coupons for Procter & Gamble products and similar promotional giveaways, and the popularity of the first print run of 10,000 copies inspired them to stick a ten-cent price tag on another issue and sell them on newsstands. The speed with which these newsstand comics sold out (despite the fact that they contained only reprints of material that had been previously available in Sunday newspapers) demonstrated that there was a future in these "funny books."

Newspaper comic strips were leased to regional papers through "distribution syndicates," which controlled the reprint rights to these strips. In order to satisfy the demand for material by comic-book publishers who couldn't secure (or didn't want to pay for) reprint rights for newspaper strips, Major Malcolm Wheeler-Nicholson hired a group of work-hungry young artists and writers and commissioned original comic material. These brand-new stories, not reprints of established newspaper strips, were published as *New Fun Comics* by National Allied Publications. The drawn, inked, lettered, and ready-for-printing comic pages that came out of Wheeler-Nicholson's studio enabled the publishers to avoid the high rates charged by the then-powerful typesetters' unions. Soon the stories that had occupied the pulp-fiction magazines were being told in graphic form, and comics featuring detective and police stories, horror, funny animals and straight gags, adventure heroes, secret agents, and crime fighters with mystical powers filled the newsstands. These ranks were joined, in 1938, by a strange visitor from another planet with powers and abilities far beyond those of mortal men.

Superman was the brainchild of Jerry Siegel and Joseph Shuster, two teenagers from Cleveland who dreamed of earning riches through the creation of a popular newspaper adventure strip. Combining attributes from two of Edgar Rice Burroughs's characters, Tarzan and John Carter of Mars, Siegel and Shuster turned the conventional science-fiction adventure story on its head. Instead of an ordinary Earthman traveling to a strange new planet (as in the stories featuring Flash Gordon or Buck Rogers), a citizen of a distant world with strange powers came to Earth. This innovation, together with the colorful uniform the hero wore (inspired, perhaps, by the costumed strongmen that performed in circuses at the time), and the then relatively novel introduction of a secret identity for the heroic adventurer made their strip so original that it was swiftly turned down by every newspaper distribution syndicate they approached. After four years of constant rejection, Siegel and Shuster were desperate enough to try to sell their Superman concept to the decidedly lesser market of comic books. There they eventually found a receptive audience in the form of Sheldon Mayer, a young editor who saw potential in the rough, early strips of Siegel and Shuster. Mayer convinced Vin Sullivan that this new character was exactly what was needed for a new comic-book title that was due at the printer but lacked a lead story. With no time to alter the strip to fit the comic-book format, panels from the two weeks of sample newspaper strips were hastily cut and pasted into a thirteen-page story. With a cover adapted from one of the panels in the strip showing Superman lifting an auto sedan over his head while crooks fled in panic, *Action Comics* # 1, cover price ten cents, appeared on newsstands in June 1938. The rest, to coin a phrase, is history.

* * *

Evolutionary biology teaches that random mutations can lead to the creation of new species. When these new species exhibit superior adaptation to a changing environment, they can quickly dominate an ecological niche. Similarly, the superhero comic book struck a resonant chord with Depression-era readers, and was an instant success. Soon the newsstands were full of superhero comics, featuring characters possessing a dazzling array of powers and abilities.

All of these new characters shared the crucial attribute of differing sufficiently from Superman, to avoid joining Fawcett publication's Captain Marvel and being sued for copyright infringement by National Publications (which owned the legal rights to Siegel and Shuster's creation). Many of these newborn heroes had a single superpower, such as superspeed (the Flash, Johnny Quick), flight (Hawkman, Black Condor), superstrength (Hourman, Captain America), or none of the above (Batman). Some of these heroes gained their superpowers via "scientific means." The Flash of the 1940s, for example, became superfast after a chemical laboratory accident in which he inhaled some "hard water."* Chemist Rex Tyler developed a pill that provided enhanced strength and speed for sixty minutes, enabling him to fight crime

---

* Water is characterized as "heavy" when the two hydrogen atoms in H_2O each have an extra neutron, while water with a large mineral content is termed "hard." Fortunately for the Golden Age Flash, water softeners were not common in the 1940s.

as Hourman. The 4F army reject Steve Rogers became the superhero Captain America, following a series of injections with a "super soldier" serum (nowadays, these would probably be described
as "steroids").

Much more common, though, was a mystical or supernatural
origin for the characters' abilities, due to the acquisition of or exposure to a magical object from some hidden corner of the globe. In
this way, as they would throughout their existence, comic books
merely reflected the popular cultural zeitgeist. For example, in the
1940s the Green Lantern was a hero who had come into possession
of a mystical lantern originally from ancient China, from which he
fashioned a ring that endowed the wearer with a wide variety of
powers but was ineffective against wood. Viewed in the cultural
context of the times, the world was a bigger place back in 1940. To
the adolescent imagination the "Far East" and the "Congo" were
still vast repositories of powerful secrets and mysterious artifacts.
When the Green Lantern character was reinvented in 1959, he was
given a new costume and origin, and the new lantern and ring became extraterrestrial artifacts. The ring's vulnerability to objects
that were colored yellow was now attributed to a chemical impurity in the ring's composition, which could not be removed without
making the ring ineffective.* Similarly, Green Lantern's colleague,
Hawkman, in 1940 was an Egyptian prince reincarnated into present times while the 1960s version of the same hero was an inter-
galactic policeman from the planet Thanagar.

This origin transition continues today. In 1962, Peter Parker
gained the powers of Spider-Man when he was bitten by a spider
that had accidentally become radioactive in a physics lab demonstration while in a 2000 reinterpretation of the same character (as
well as in the 2002 film version), the fateful bite was from a genetically engineered superspider that escapes during a molecular-
biology lab demonstration. Thus the one constant would appear to
be that the creation of the superhero is a way of binding the cultural anxiety of the day, whether of the "distant other" in the
1940s, radioactivity in the 1960s, or genetic manipulation today.

The original incarnations of various superheroes in the late
1930s and 1940s were products of their time and reflected life

* Hence an effective anti–Green Lantern weapon, regardless of the period,
would be a yellow, wooden baseball bat.

during the Great Depression and World War II. After the war, soldiers who had acquired a comic-book-reading habit while overseas continued to buy comics after returning to the States, and certain publishers catered to this older audience with adult-themed stories featuring more graphic violence. Some of the young comic-book writers and artists had also been drafted into the armed services, and their wartime experiences resulted in a more serious, and in some cases darker, tone to their postwar work. From their inception, comic books were intended to attract a younger readership. In 1945 Maxwell Gaines ended his association with National Comics and started a new publishing firm called Educational Comics, printing such titles as *Picture Stories from Science, Picture Stories from American History,* and *Picture Stories from the Bible.* After his untimely passing in 1947, his son William Gaines changed the firm's name to Entertaining Comics (EC) and shifted their inventory to such comics as *Tales from the Crypt, Crime SuspenStories, Weird Science-Fantasy,* and *The Vault of Horror.* These comics were neither suitable nor intended for the same audience as Captain Marvel. It was just a matter of time before someone noticed and complained.

Dr. Fredric Wertham's 1953 best-selling book, *Seduction of the Innocent,* forcefully argued that such lurid stories corrupted the minds of young children, leading them directly to careers as juvenile delinquents. In a cycle that appears to repeat itself in every generation, there was a growing concern among parents and authority figures in the post–World War II era over the coarsening effects of popular culture on the attitudes and mores of teenagers. The U.S. Senate Subcommittee on Juvenile Delinquency, headed by the ambitious Sen. Estes Kefauver, held hearings on the connection between comic books and teenage crime. Initially the committee intended to focus solely on horror and crime comics, but Wertham, a consultant to the subcommittee, brought superhero comics to the senators' attention. Seeking to avoid the imposition of federal oversight and regulation, the major comic-book publishers created a self-regulatory agency called the Comics Code Authority (CCA). The publishers developed a set of rules governing acceptable comic-book content, with explicit instructions that gore, lewdness, drug use, zombies, and vampires were prohibited in any comic book bearing the Comics Code Authority seal of approval on its cover. Many of the guidelines created by the CCA seemed designed solely to ensure the destruction of nearly the entire EC line of comics (the only survivor being a relatively new comic satire magazine by the name of *MAD*). All comic-book stories had to be submitted to the CCA (whose staff were funded by the publishers) for approval before being published, similar to the current Ratings Board that vets motion pictures.

While it played an important role in the 1950s and 1960s in convincing parents that comic books could be viewed as "wholesome" children's entertainment, the influence of the Comics Code Authority has waned over the years as the average age of the typical comic-book reader has increased. This is reflected in the decreasing size of the CCA seal on comic-book covers. In 1964 the seal took up an area roughly that of postage stamp, or two thirds of a square inch (as it was a prominent marketing tool to convince parents that the story contained within was acceptable for their children) while it was less than one quarter of a square inch in 1984 and is a barely detectible one tenth of a square inch in 2004 (for DC Comics; Marvel Comics quit its participation in the CCA in 2001 and employs a self-determined labeling system roughly equivalent to the PG, PG-13, and R ratings used by movies).

Declining sales from the loss of a major distribution network and the competition from television led to the near collapse of the comic-book industry, and from 1953 to 1956, only about a half-dozen superhero comics continued to be published, a dramatic reduction from the 130 different superhero titles available at newsstands at the peak of the Golden Age. Funny animal stories, westerns, and young romance comics were safer alternatives for the few companies that persevered in publishing comics during this period.

In 1956 National comics decided to test the superhero waters with the reintroduction of the Flash in *Showcase* # 4. The sales figures for each issue of *Showcase* that featured the Flash indicated that the market for superheroes had returned, and over the next few years National brought back new versions of the Green Lantern, the Atom, Hawkman, and others. The Silver Age of superhero comic books had begun, and superheroes have remained a mainstay of comic books ever since.

From the very beginning in *Showcase* # 4, examples of correctly applied physics principles appeared in these stories. With the launching of the Soviet satellite Sputnik in 1956 at the height of the Cold War, there was considerable anxiety over the quality of science education that American schoolchildren were receiving. The Comics Code Authority seal on their covers guaranteed that the superhero comic-book stories were not harmful for young readers, and the inclusion of science concepts may have convinced some that there was a net positive benefit to these four-colored adventures.

In addition to employing accurate science, comics from the Silver Age often had scholarly nuggets from other learned disciplines buried within their stories. For example, the plot of "The Adventure of the Cancelled Birthday" in *The Atom* # 21 (written by Gardner Fox, who was both a lawyer and a writer for science-fiction pulp magazines) revolved around the obscure fact that in 1752, when Great Britain adopted the Gregorian calendar to replace the Julian calendar, eleven days were omitted during the transition. That is, September 2, 1752, was followed the next day by September 14, in order to regularize the British calendar with other parts of Europe. (Britons, distrustful of their government and believing that they were being cheated, rioted with the rallying cry, "Give us back our eleven days!")

Young readers were thereby introduced through this superhero comic to facts and historical figures not typically covered in their history classes. Two issues later the letters column in *The Atom* printed a complaint from one such fan, arguing over the poor choice of historical characters, such as the obscure Justice Fielding. The editor of *Atom* comics, Julius "Julie" Schwartz, responsible for reintroducing the Flash in 1956, defended the story in the letters column, pointing out that it was high time that the reader become acquainted, as had the Atom, with Henry Fielding, the author of *Tom Jones*.

Even if they were not woven into the plot, bits of historical or scientific trivia would occasionally pop up in comic-book adventures through the appearance of a caption box containing a fact that had no direct bearing on the story at hand. For example, in the *Brave and the Bold* # 28 featuring the first appearance of the alliance of National Comics superheroes forming the Justice League of America, Aquaman swims by a puffer fish and has a brief conversation with him using his "fish telepathy." The puffer fish relates some crucial information gleaned while floating on the surface of the ocean. A caption in this panel informs us that: "by swallowing air into a special sac beneath its throat, the puffer fish becomes inflated like a football—whereupon it rises to the surface and floats upside down."

Why take the time to include these educational captions? They may have been a consequence of the habits of the former pulp-fiction writers penning these tales. Prior to editing comic books at National, Mort Weisinger and Julie Schwartz, lifelong fans of science fiction, had been literary agents for science-fiction and fantasy writers, including Ray Bradbury, Robert Bloch, and H. P. Lovecraft. Certain comic-book writers had previously made their living as pulp science-fiction writers. As such, they were walking storehouses of obscure historical and natural knowledge. The Hugo Award winner Alfred Bester, author of science-fiction classics *The Demolished Man* and *The Stars, My Destination*, also wrote comics during the 1940s and penned the original Green Lantern oath. In an autobiographical essay, Bester tells of spending hours browsing through reference books in the New York Public Library, searching for odd historical tidbits around which he could construct a story. Knowing a lot of trivia could also help these pulp-fiction writers' financial bottom line, as these authors were paid by the word. Consequently they would frequently pad their work with all sorts of barely relevant tangents, as reflected in this joke:

Q: How many pulp-fiction writers does it take to change a lightbulb?

A: The history of the lightbulb is a long and interesting tale, beginning in 1879 in the quiet town of Menlo Park, New Jersey, and continuing on to the present day. . . .

While the Silver Age comic-book writers may have had an economic incentive to be verbose, it is also likely that they were motivated by considerations of self-preservation to inject educational elements into their stories. As mentioned above, the introduction of science facts and principles into these stories may have stemmed from a genuine desire on the part of the writers and editors to educate, or perhaps simply from a survival instinct to avoid any further congressional attention.

A PHYSICIST READS A COMIC BOOK

Reading classic and contemporary superhero comic books now, with the benefit of a Ph.D. in physics, I have found many examples of the correct description and application of physics concepts in superhero comic books. Of course, nearly without exception, the use of superpowers themselves involves direct violations of the known laws of physics, requiring a deliberate and willful suspension of disbelief. However, many comics needed only a single "miracle exception"—one thing you have to buy into to make the superhero plausible—and the rest that follows as the hero and villain square off would be consistent with the principles of science. While the intent of these stories has always primarily been to entertain, if at the same time the reader was also educated, either deliberately or accidentally, this was a happy bonus.

It is these happy bonuses, such as the one illustrated in fig. 2, that I wish to consider here. In this book I'll present an overview of certain scientific principles, using examples of their correct application as found in comic books. I will describe characters and situations that illuminate various physics concepts, rather than systematically considering the physics underlying an array of superheroes. (Consequently it is conceivable that your favorite superhero may not be discussed. I know that several of my own favorites didn't make the cut.) By the end of this book the reader will have been exposed to the key concepts in an introductory physics class, with a little bit of upper-level quantum mechanics and solid-state physics thrown in for fun. By examining the physical principles underlying certain comic-book adventures, we will at the same time gain an understanding of the mechanisms behind many real-world practical applications, from television to stellar nucleosynthesis of the elements to telephones.

I will focus primarily but not exclusively on the Silver Age period in comic-book history (from the reintroduction of the Flash in *Showcase* # 4 in 1956 to the death of Gwen Stacy in *The Amazing Spider-Man* # 121 in 1973) because the writers of this period made more of an effort than those in the Golden Age to incorporate scientific principles into their stories. In addition, the Silver Age characters have demonstrated lasting popularity, and their iconic status will make it easier to refer to their exploits without forcing the reader to constantly consult the back-issue bins of their local comic-book shop to uncover their backstory. It is all too easy to find flaws and errors in the science referenced in comic-book stories, and this is not the aim of this book. In addition to being unsporting and uncharitable (as should be obvious, these stories were never intended to function as science textbooks, despite the occasional student's attempts at surreptitious substitution), it is more difficult to make a point when the only illustrative examples are negative ones. Nevertheless, sometimes we will find that some scenes in comic books are simply not physically plausible, even granting a "miracle exception."

Before I begin I would like to say a few words about a common misconception concerning physicists. Despite the impression gleaned from popular movies, being a physicist does not require an encyclopedic knowledge of equations and fundamental constants, coupled with the ability to perform complex arithmetic in one's head with robotic speed and precision. Physics is not about having memorized all the answers, but rather about asking the right questions. For when the right question is posed of a phenomenon, either the answer becomes clear, or at least the manner as to how one should go about obtaining the answer is revealed.

To illustrate that asking one right question can be more important than a bushel full of correct answers, consider the simple physics experiment of tossing a ball in an arc. There are many questions we may ask, such as: How high does the ball travel? How far to the right does it move? How long is it in the air? How fast is it going? What is the geometric shape of its path? However, I would argue that there is *one* simple question that implies all of the above questions and gets at the heart of the issues concerning the ball's motion. That one single question is the following: **Does the ball have any choice?** If the ball does not have any choice in its motion, if it lacks free will, then its trajectory is completely determined by forces external to itself. Once we determine the nature of these forces and how they influence the ball's motion, we may then calculate the path of the ball for a given initial velocity imparted by the thrower. This calculated trajectory would then contain any and all information we may desire regarding how high the ball rises, how far it moves, its time in flight, what its velocity is, and so on. If we then repeat the toss with exactly the same initial position and velocity as before, then the ball *must* exactly and faithfully trace out the calculated trajectory, for the ball does not have any choice in the matter.

This is the beauty and attraction of physics, at least for those of us lucky enough to make our living from its study. The promise and potential is that if we can determine the forces acting on an object and how these forces influence the object's motion, we will then be able to predict the development of future events. By performing careful experiments, these predictions can be empirically tested and, if correct, confirm our understanding of how nature operates. On the other hand, if the experiment contradicts our model (a far more likely outcome initially), we modify our equations and try again, using the failed test as an important clue as to what was missing from the initial calculation.* In this way our understanding of nature progresses until we have a valid model, which is then termed a theory. To dismiss any idea that makes it through this exhaustive vetting as "just a theory" is equivalent to describing the Hope diamond as "just a crystal."

Scientific knowledge only comes at the price of increased doubt: The more we learn, the more clearly we see all that remains uncertain. Doubt is to be embraced in science, for the only answers we can trust are those that survive the crucible of questioning and experimental testing. I hope to share with you in this book the true pleasure that comes from seeing how the asking of a few key questions can lead to a wealth of answers about the world we live in.

I begin, as do all standard textbooks in freshman physics, with the fundamental laws of motion as first described by Isaac Newton. Fitting such an original and profound contribution to modern

---

* The accumulated body of knowledge about the world is now so vast that physicists are able to make continued progress only by specializing in either experimental or theoretical research. Experimentalists work in laboratories and carry out measurements while theoreticians perform calculations and computer simulations. I am an experimentalist, while Stephen Hawking is a theorist (the differences begin there). One of the last physicists who truly excelled at *both* experimental and theoretical research was Enrico Fermi.

thought, our first comic-book example involves an equally semi-nal contribution to Western civilization. I refer, of course, to the first true comic-book superhero, faster than a speeding bullet, more powerful than a locomotive, and, most relevant to our next dis-cussion, able to leap tall buildings in a single bound.

SECTION
1

MECHANICS

UP, UP, AND AWAY—

FORCES AND MOTION

JERRY SIEGEL AND JOSEPH SHUSTER'S original conception of Superman was of a pulp action hero with a liberal dose of science fiction added to lend an air of plausibility for their hero's great strength. As described in *Superman* # 1, Jor-El, a scientist on the distant planet Krypton, discovers that his world is about to explode and kill its entire population. Possessing only a small prototype rocket ship, he and his wife elect to save their infant son Kal-El, sending him to Earth so that he will not share their fate.* After traveling great distances through the vastness of space, the rocket crash-lands on Earth with its sole passenger none the worse for wear. Discovered by the childless Kansas farmers the Kents, Kal-El is immediately given up to an orphanage. Later, driven by guilt, the Kents return to the orphanage (where the superbaby has been wreaking havoc) whereupon they adopt Kal-El/name him Clark, and raise him as their own human son. As Kal-Clark Kent grows into adulthood, he develops a series of extraordinary abilities with which he fights the never-ending battle for Truth, Justice, and the American Way.

The source of Superman's powers on Earth was originally credited to his Kryptonian heritage, specifically the fact that his home planet had a far stronger gravity than Earth's. For example, the

---

* As a father myself, I can certainly empathize with Jor-El. Many are the times I've been tempted to place my own kids in a rocket ship and send them off into deep space.

moon's much smaller size compared to Earth results in a weaker gravitational field, so objects on the moon weigh less than they do on Earth. Consequently, an Earthman whose muscles and bones are adapted to Earth's gravity is able to leap far greater distances and lift more massive objects on the lunar surface. Similarly Superman's great strength ("more powerful than a locomotive") and tougher skin ("nothing less than a bursting shell" could pierce it) resulted from his relocating to a planet with a far weaker gravity than Krypton's. Even though Superman was sent to Earth as an infant, presumably his Kryptonian DNA encoded for the development of muscles and bones suited to a stronger gravitational field.

By the late 1940s, Superman would gain the power of true flight, able to choose and alter his trajectory after leaving the ground. At this point Superman can be considered to have gained free will over the laws of physics. Over time he acquired a host of other abilities that could not be reasonably accounted for by the stronger gravity of his home planet. These powers included various visions (heat-, X-ray-, and others), super-hearing, super-breath, and even super-hypnotism.*

The origin of Superman's powers was subsequently revised in *Action Comics* # 262, to claim that Superman's fantastic abilities derived from the fact that the Earth orbited a yellow sun, as opposed to the red sun of Krypton. The color of a sun is a function of both its surface temperature and the atmosphere through which it is viewed. The blue portion of the solar spectrum is strongly scattered by the atmosphere, which is why the sky looks blue. Viewed straight on, our sun appears yellow because the atmosphere is also more absorbing toward the blue end of the spectrum, except at dawn or sunset when the position of the sun is low on the horizon and sunlight must travel a greater distance through the atmosphere. Nearly all wavelengths are then absorbed except for lower-energy red light, which gives sunsets their characteristic hues (the

---

* This last power was introduced to explain why only a simple pair of eye-glasses created such a perfect disguise that no one ever realized that mild-mannered reporter Clark Kent and the world-famous Superman were the same person. As described in *Superman* # 330, Superman apparently subconsciously hypnotizes everyone who sees him into believing that his face is markedly different from Clark Kent's.

greater number of particulates in the air at the end of the day compared with the beginning also contributes to the difference in shading between sunset and dawn). These spectral features are for the most part independent of the chemical composition of the gases making up the Earth's atmosphere. There is no physical mechanism by which a shift in the primary wavelength of sunlight from yellow (a wavelength of 570 nanometers, or 570 billionths of a meter) to red light (650 nanometers) would endow someone with the ability to bend steel in his bare hands. Consequently at this stage in his history, *Superman* ceased being a science-fiction strip, and became a comic book about a fantasy hero. Changing a superhero's origin in order to accommodate new powers or circumstances occurs so frequently in comic books that comic-book fans have coined a term, "retconning," to describe this retroactive continuity repair.

Interestingly, Superman's foes went through a similar evolution around this same time. In the early years of *Action* and *Superman* comics, Siegel and Shuster gave voice to the revenge fantasies of their young and economically disadvantaged Depression-era readers. Superman first used his powers to fight corrupt slumlords, coal-mine owners, munitions manufacturers, and Washington lobbyists. In his very first story he psychologically tormented a lobbyist by holding him as they both fell from a tall building. At this early stage of his career, the story lines indicated that only a few people knew of Superman's existence, and the lobbyist believed that the fall would be fatal. He willingly divulged the information Superman was after rather than risk another such fall. By the 1950s, in addition to selling millions of comics per month, Superman had become a star of radio serials, movie shorts (both animated and live action), and a popular television program. Around this time his adversaries morphed into criminal masterminds with colorful personas and costumes, such as the Toyman, the Prankster, and Lex Luthor, whose schemes for grand larceny or world (or in the case of Brainiac, galactic) domination Superman was able to foil while keeping the corporate power structure safely undisturbed. As befits the escalating capabilities of the villains he faced, Superman entered a superpower arms race, eventually growing so powerful that it became difficult for writers to concoct credible threats to challenge his near godlike abilities. Radioactive fragments of his home planet, known as Kryptonite, became a frequent device to extend any given story beyond the first page of the comic.*

It is the simpler, original Superman of the Golden Age, the last son of Krypton, that I wish to consider here.

THE PHYSICS OF JUMPING AND ALL OTHER MOTION

In the first years of his comic-book history, Superman was unable to fly but could simply "leap tall buildings in a single bound," thanks to Earth's weaker gravity.

Well, how high could he leap? According to his origin story in *Superman* # 1, Superman's range was about one eighth of a mile, or 660 feet. Assuming he could jump this high straight up, this is approximately equivalent to the height of a thirty- to forty-story building, which in 1938 would be considered quite tall. So our question can be rephrased as: What initial velocity would Superman need, lifting off from the sidewalk, so that he would vertically rise 660 feet?

Whether we describe the trajectory of a leaping man of steel or of the tossed ball from our earlier example, we must begin with the three laws of motion as first described by Isaac Newton in the mid-1600s. These laws are frequently expressed as: (1) an object at rest remains at rest, or if moving keeps moving in a straight line if no external forces act upon it; (2) if an external force *is* applied, the object's motion will change in either magnitude or direction, and the rate of change of the motion (its acceleration) when multiplied by the object's mass is equal to the applied force; and (3) for every force applied to an object there is an equal and opposite force

---

* How a radioactive element from his native planet would affect Superman so strongly, while he remained immune to radioactive isotopes found on Earth, is more an issue of literary necessity rather than physical plausibility. Kryptonite was first introduced in the radio serial *Adventures of Superman* in 1943, when the overworked radio actor portraying the Man of Steel wanted a vacation. The radio scriptwriters created Superman's mineral nemesis so that another actor could portray the stricken superhero by groaning into the microphone. Several years later the comic-book writers adopted and adapted this creative device, and a rainbow of Kryptonite (green, red, gold, silver, and others) with a broad range of effects on Superman was introduced into the comic-book universe.

exerted back by the object. The first two laws can be expressed succinctly through one simple mathematical equation:

FORCE = (MASS) X (ACCELERATION)

That is, the force **F** applied to an object is equal to the resulting rate of change in the object's velocity (its acceleration **a**) when multiplied by the object's mass **m**, or **F = ma**.

Acceleration is a measure of the rate of change of the velocity of an object. A car starting from rest (velocity = 0) and accelerating to 60 mph would have a change in velocity of 60 mph − 0 mph = 60 mph. The acceleration is then given by dividing this change in velocity by the time needed to make this change. The longer the time, the lower the acceleration needed for a given change in speed. An automobile speeding up from 0 to 60 mph in six seconds will have a much larger acceleration than if it does so in six hours or six days. The final speed will be the same for all three cases, namely 60 mph, but the accelerations will be radically different owing to the different times needed to affect this change in velocity. From Newton's **F = ma**, the force needed to create the former, faster acceleration is obviously much larger than for the latter, slower case.

When the acceleration is zero, there is no change in the motion. In that case a moving object keeps moving in a straight line, or if sitting still, remains so. From the expression **F = ma**, when **a = 0**, then the force **F = 0**, which is the whole point of Newton's first law of motion.

While this may be straightforward from a mathematical point of view, from a common-sense perspective it is nothing short of revolutionary. Newton is saying (correctly) that if an object is moving, and there is no outside force acting on it, then the object will simply continue moving in a straight line. However, you and I, and Isaac Newton for that matter, have *never* seen this occur! Our everyday experiences tell us that to keep something moving, we must always keep pulling it or pushing it with an external force. A car in motion does *not* remain in motion, unless we keep pressing the accelerator pedal, which ultimately provides a force. Of course, the reason that moving objects slow down and come to rest when we stop pushing or pulling them is that there are forces of friction and air resistance that oppose the object's motion. Just because we stop pulling or pushing does not mean, in the real world, that there are no forces acting on the object. There's nothing wrong with Newton's laws—we just have to make sure we account for friction and air resistance when applying them. It is these unseen "drag forces" that we must overcome in order to maintain uniform motion. Once our pulling or pushing exactly balances the friction or air drag, then the *net* force on the object *is* zero, and the object will then continue in straight-line motion. Increasing the push or pull further will yield a net nonzero force in the direction of our push or pull. In this case, there will be an acceleration proportional to the net force. The constant of proportionality connecting the force to the acceleration is the mass, **m**, reflecting how much the object resists changing its motion.

It is worth pointing out here that mass is not the same as weight. "Weight" is another term for "force on an object due to gravity." Mass, on the other hand, is a measure of how much stuff ("atoms" for you specialists) an object contains. The mass of the atoms in an object is what gives it its "inertia," a fancy term to describe its resistance to change when a force is applied. Even in outer space, an object's mass is the same as on the Earth's surface, because the number of atoms it contains does not change. An object in outer space may be "weightless," in that it is subject to a negligible attractive force from nearby planets, but it still resists changes in motion, due to its mass. A space-walking astronaut in deep space cannot just pick up and toss the space station around (assuming she had a platform on which to stand), even though the station and everyone on it is "weightless." The mass of the space station is so large that the force the astronaut's muscles can apply produces only a negligible acceleration.

For objects on the Earth's surface (or that of any other planet, for that matter), the acceleration due to gravity is represented by the letter **g** (we'll discuss this more in a moment). The force that gravity exerts on the object of mass **m** is then referred to as its Weight. That is, **Weight = (mass) × (acceleration due to gravity)** or **W = mg**, which is just a restatement of $F = ma$ when $a = g$. Mass is an intrinsic property of any object, and is measured in kilograms in the metric system, while Weight represents the force exerted on the object due to gravity, and is measured in pounds in the United States. In Europe, Weight is commonly expressed in units of kilograms, which is not strictly correct, but easier to say than "kilogram–meter/sec," the unit of force in the metric system (also known as a "Newton").

When something weighing one Newton in the metric system is compared to an object weighing one pound in the United States, the conversion ratio is *1 kilogram is equivalent to 2.2 pounds.* I say "equivalent" and not "equal" because a pound is a unit of force, while kilograms measure mass. An object will weigh less than 2.2 pounds on the moon and more than 2.2 pounds on Jupiter, but its mass will always be 1 kilogram. When calculating forces in the metric system, we'll stick with kg-meter/sec^2 rather than "Newtons," in order to remind ourselves that *any* force can always be described by $\mathbf{F = ma}$.

To recap, Superman's mass at any given moment is a constant, because it reflects how many atoms are in his body. His weight, however, is a function of the gravitational attraction between him and whatever large mass he is standing on. Superman has a larger weight on the surface of Jupiter, or a lesser weight on the Moon, compared to his weight on Earth, but his mass remains unchanged. The gravitational attraction of a planet or moon decreases the farther away one moves from the planet, though technically it is never exactly zero unless one were infinitely far from the planet. It is tempting to equate mass with weight, and easy to do so when dealing only with objects on Earth for which the acceleration due to gravity is always the same. As we will soon be comparing Superman's weight on Krypton to that on Earth, we will resist this temptation.

Finally, the third law of motion simply makes explicit the commonsense notion that when you press on something, that thing presses back on you. This is sometimes expressed as "For every action, there is an equal and opposite reaction." You can only support yourself by leaning on the wall if the wall resists you—that is, pushes back with an equal and opposite force. If the force were not exactly equal and in the opposite direction, then there would be a *net* nonzero force, which would lead to an acceleration and you crashing into the wall. When the astronaut mentioned above pushes on the space station, the force her muscles exert provides a very small acceleration to the station, but the station pushes back on her, and her acceleration is much larger (since her mass is much smaller).

Imagine Superman and the Hulk holding bathroom scales against each other (which are simply devices to measure a force, namely your weight due to gravity). When they press against each other's scale, no matter how hard Superman pushes on the left, if they remain stationary, then the Hulk's scale on the right will read exactly the same force. Moreover, no matter how hard Superman is pushing, his scale will read zero pounds of force if the Hulk offers no resistance and just moves his scale out of the way and steps aside.* *Forces always come in pairs*, and you cannot push or pull on something unless it pushes or pulls back. When you stand on the sidewalk, your feet exert a force on the ground due to gravity pulling you toward the center of the Earth. People on the opposite side of the planet do not fall off, because gravity pulls everyone in toward the center of the planet, regardless of where they are located. You do not accelerate while standing; the ground provides an equal and opposite force exactly equal to your weight. During the brief moment when Superman jumps, his legs exert a force greater than just his normal standing weight. Because forces come in pairs, his pushing down on the pavement causes the pavement to push back on him. Thus he experiences an upward force lifting him up and away.

And that's it—all of Newton's laws of motion can be summarized in two simple ideas: that any change in motion can only result from a external force ($F = ma$), and that forces always come in pairs. This will be all we need to describe all motions, from the simple to the complex, from a tossed ball to the orbits of the planets. In fact, we already have enough physics in hand to figure out the initial velocity Superman needs to leap a tall building.

IN A SINGLE BOUND

Superman starts off with some large initial velocity (fig. 4). At the top of his leap, a height $h = 660$ feet above the ground, his final velocity must be zero, or else this wouldn't be the highest point of his jump, and he would in fact keep rising. The reason Superman slows down is that an external force, namely gravity, acts on him. This force acts downward, toward the surface of the Earth, and opposes his rise. Hence the acceleration is actually a deceleration,

---

* The Hulk is brighter than everyone gives him credit for (his alter ego *is* a physicist, after all).

SUPERMAN LAUNCHES HIMSELF UP ALONG THE SIDE OF THE BUILDING IN A GREAT LEAP !

Fig. 4. *Panel from* Superman # 1 *(June 1939) showing Superman in the process of leaping a . . . well, you know.*

slowing him down, until at 660 feet, he comes to rest. Imagine ice-skating into a strong, stiff wind. Initially you push off from the ice and start moving quickly into the wind. But the wind provides a steady force opposing your motion. If you do not push off again, then this steady wind slows you down until you are no longer moving and you come to rest. But the wind is still pushing you, so you still have an acceleration and now start sliding backward the way you came, with the wind. By the time you reach your initial starting position, you are moving as fast as when you began, only now in the opposite direction. This constant wind in the horizontal direction affects you as an ice-skater the same way gravity acts on Superman as he jumps. The force of gravity is the same at the start, middle, and highest point of his leap. Since $F = ma$, his acceleration is the same at all times as well. In order to determine what starting speed Superman needs to jump 660 feet, we have to figure out how his velocity changes in the presence of a uniform, constant acceleration g in the downward direction.

As common sense would indicate, the higher one wishes to leap, the faster the liftoff velocity must be. How, exactly, are the starting speed and final height connected? Well, when you take a trip, the distance you travel is just the product of your average speed and the length of time of the trip. After driving for an hour at an average speed of 60 mph, you are 60 miles from your starting point. Because we don't know how long Superman's leap lasts, but only his final height of $h - 660$ feet, we perform some algebraic manipulation of definition that acceleration is the change in speed over time and that velocity is the change in distance over time. When the dust settles we find that the relationship between Superman's initial velocity v and the final height h of his leap is $v \times v = v^2 = 2gh$. That is, the height Superman is able to jump depends on the *square* of his liftoff velocity, so if his starting speed is doubled, he rises a distance four times higher.

Why does the height that Superman can leap depend on the square of his starting speed? Because the height of his jump is given by his speed multiplied by his time rising in the air, and the time he spends rising *also* depends on his initial velocity. When you slam on your auto's brakes, the faster you were driving, the longer it takes to come to a full stop. Similarly, the faster Superman is going at the beginning of his jump, the longer it takes gravity to slow him down to a speed of zero (which corresponds to the top of his jump). Using the fact that the (experimentally measured) acceleration due to gravity g is 32 feet per second per second (that is, an object dropped with zero initial velocity has a speed of 32 feet/sec after the first second, 64 feet/sec after the next second, and so on) the expression $v^2 = 2gh$ tells us that Superman's initial velocity must be 205 feet/sec in order to leap a height of 660 feet. That's equivalent to 140 miles per hour! Right away, we can see why we puny Earthlings are unable to jump over skyscrapers, and why I'm lucky to be able to leap a trash can in a single bound.

In the above argument we have used Superman's average speed, which is simply the sum of his starting speed (v) and his final speed (zero) divided by two. In this case his average speed is $v/2$, which is where the factor of two in front of the gh in $v^2 = 2gh$ came from. In reality, both Superman's velocity and position are constantly decreasing and increasing, respectively, as he rises. To deal with continuously changing quantities, one should employ calculus (don't worry, we won't) while so far we have only made use of algebra. In order to apply the laws of motion that he described, Isaac Newton had to first *invent* calculus before he could carry out his calculations, which certainly puts our difficulties with mathematics into some perspective. Fortunately for us, in this situation, the rigorous, formally correct expression found using calculus turns out to be exactly the same as the one obtained using relatively simpler arguments, that is $v^2 = 2gh$.

How does Superman achieve this initial velocity of more than 200 feet/sec? As illustrated in fig. 5, he does it through a mechanical process that physicists term "jumping." Superman crouches down and applies a large force to the ground, causing the ground to push back (since forces come in pairs, according to Newton's third law). As one would expect, it takes a large force in order to jump up with a starting speed of 140 mph. To find exactly how large a force is needed, we make use of Newton's second law of motion, $F = ma$—that is, Force is equal to mass multiplied by acceleration. If Superman weighs 220 pounds on Earth, he would have a mass of 100 kilograms. So to find the force, we have to figure out his acceleration when he goes from standing still to jumping with a speed of 140 mph. Recall that the acceleration describes the change in velocity divided by the time during which the speed changes. If the time Superman spends pushing on the ground using his leg muscles is 1/4 second, then his acceleration will be the change in speed of 200 feet/sec divided by the time of 1/4 second, or 800 feet/sec^2 (approximately 250 meters/sec^2 in the metric system, because a meter is roughly 39 inches). This acceleration would correspond to an automobile going from 0 to 60 mph in a tenth of a second. Superman's acceleration results from the force applied by his leg muscles to get him airborne. The point of $F = ma$ is that for any change in motion, there must be an applied force

Fig. 5. *Panels from* Action *comics # 23, describing in some detail the process by which Superman is able to achieve the high initial velocities necessary for his mighty leaps.*

and the bigger the change, the bigger the force. If Superman has a mass of 100 kilograms, then the force needed to enable him to vertically leap 660 feet is $\mathbf{F} = \mathbf{ma} = (100 \text{ kilograms}) \times (250 \text{ meters/sec}^2) = 25,000$ kilograms meters/sec^2, or about 5,600 pounds.

Is it reasonable that Superman's leg muscles could provide a force of 5,600 pounds? Why not, if Krypton's gravity is stronger than Earth's, and his leg muscles are able to support his weight on Krypton? We calculated that when making his greatest leap, Superman's legs must provide a force of 5,600 pounds. Suppose that this is 70 percent larger than the force his legs supply while simply standing still, supporting his weight on Krypton. (This is being generous, as when most people jump they can only apply a force approximately equal to their standing weight.) In this case, Superman on his home planet would weigh 3,300 pounds. His weight on Krypton is determined by his mass and the acceleration due to gravity on Krypton. We assumed that Superman's mass is 100 kilograms, and this is his mass regardless of which planet he happens to stand on. If Superman weighs 220 pounds on Earth and nearly 3,300 pounds on Krypton, then the acceleration due to gravity on Krypton must have been 15 times larger than that on Earth.

So, just by knowing that $\mathbf{F} = \mathbf{ma}$, making use of the definitions "distance = speed × time" and "acceleration is the change in speed over time," and the experimental observation that Superman can "leap a tall building in a single bound," we have figured out that **the gravity on Krypton must have been 15 times greater than on Earth.**

Congratulations. You've just done a physics calculation!

DECONSTRUCTING KRYPTON—

NEWTON'S LAW OF GRAVITY

NOW THAT WE HAVE DETERMINED THAT in order for Superman to leap a tall building he must have come from a planet with a gravitational attraction 15 times that of Earth, we next ask: how would we go about building such a planet? To answer this, we must understand the nature of a planet's gravitational pull, and here again we rely on Newton's genius. What follows involves more math, but bear with me for a moment. There's a beautiful payoff in a few pages that explains the connection between Newton's apple and gravity.

As if describing the laws of motion previously discussed and inventing calculus weren't enough, Isaac Newton also elucidated the nature of the force two objects exert on each other owing to their gravitational attraction. In order to account for the orbits of the planets, Newton concluded that the force due to gravity between two masses (let's call them **Mass 1** and **Mass 2**) separated by a distance **d** is given by

$$\text{FORCE DUE TO GRAVITY} = (G) \times \frac{[(\text{MASS 1}) \times (\text{MASS 2})]}{(\text{DISTANCE})^2}$$

where **G** is the universal gravitational constant. This expression describes the gravitational attraction between *any* two masses, whether between the Earth and the Sun, the Earth and the Moon or between the Earth and Superman. If one mass is the Earth and the other mass is Superman, then the distance between them is the radius of the Earth (the distance from the center of the Earth to the surface, upon which the Man of Steel is standing). For a spherically symmetric distribution of mass, such as a planet, the attractive force behaves as if all of the planet's mass is concentrated at a single point at the planet's core. This is why we can use the radius of the Earth as the distance in Newton's equation separating the two masses (Earth and Superman). The force is just the gravitational pull that Superman (as well as every other person) feels. Using the mass of Superman (100 kilograms), the mass of the Earth, and the distance between Superman and the center of the Earth (the radius of the Earth), along with the measured value of the gravitational constant in the previous equation gives the force **F** between Superman and the Earth to be **F** = 220 pounds.

But this is just Superman's weight on Earth, which is measured when he steps on a bathroom scale on Earth. The cool thing is that these two expressions for the gravitational force on Superman are the same thing! Comparing the two expressions for **Superman's weight = (Mass 1) × g** and the **force of gravity = (Mass 1) × [(G × Mass 2)/(distance)²]**, since the forces are the same and Superman's **Mass 1** = 100 kg is the same, then the quantities multiplying **Mass 1** must be the same; that is, the acceleration due to gravity **g** is equal to **(G × Mass 2)/d²**. Substituting the mass of the Earth for **Mass 2** and the radius of the Earth for **d** in this expression gives us **g** = 10 meter/sec² = 32 feet/sec².

The beauty of Newton's formula for gravity is that it tells us *why* the acceleration due to gravity has the value it does. For the same object on the surface of the moon, which has both a smaller mass and radius, the acceleration due to gravity is calculated to be only 5.3 feet/sec²—about one sixth as large as on Earth.

This is the true meaning of the story of Isaac Newton and the apple. It certainly wasn't the case that in 1665 Newton saw an apple fall from a tree and suddenly realized that gravity existed, nor did he see an apple fall and immediately write down **F = G (m₁ m₂)/(d)²**. Rather, Newton's brilliant insight in the seventeenth century was that the exact same force that pulled the apple toward the Earth pulled the moon toward the Earth, thereby connecting the terrestrial with the celestial. In order for the moon to stay in a circular orbit around the Earth, a force has to pull on it in order to constantly change its direction, keeping it in a closed orbit.

Remember Newton's second law of **F = ma:** If there's no force, there's no change in the motion. When you tie a string to a bucket and swing it in a horizontal circle, you must continually pull on the string. If the tension in the string doesn't change, then the bucket stays in uniform circular motion. The tension in the string is not acting in the direction that the bucket is moving; consequently it can only change its direction but not its speed. The moment you let go of the string, the bucket will fly away from you.

Back to the case of the moon. If there were no gravity, no force acting on it, then the moon would travel in a straight line right past the Earth. If there were gravity but the moon were stationary, then it would be pulled down and crash into our planet. The moon's distance from the Earth and its speed are such that they exactly balance the gravitational pull, so that it remains in a stable circular orbit. The moon does not fly away from us, because it is pulled by the Earth's gravity, causing it to "fall" toward the Earth, while its speed is great enough to keep the moon from being pulled any closer to us. The same force that causes the moon to "fall" in a circular orbit around the Earth, and causes the Earth to "fall" in an elliptical orbit around the sun, causes the apple to fall toward the Earth from the tree. And that same gravitational force causes Superman to slow in his ascent once he leaps, until he reaches the top of a tall skyscraper. Once we know that in order to make such a powerful leap his body had to be adapted to an environment where the acceleration due to gravity is 15 times greater than on Earth, that same gravitational force informs us about Krypton's geology.

One consequence of Newton's law of gravitation—which states that as the distance between two objects increases, the gravitational pull between them becomes weaker by the square of their separation—is that all planets are round. A sphere has a volume that grows with the cube of the radius of the orb, while its surface area increases with the square of the radius. This balancing of the square of the radius for the surface area with the inverse square of the gravitational force leads to a sphere being the only stable form that a large gravitational mass can maintain. In fact, to address the astrophysical question of what distinguishes a very large asteroid from a very small planet, one answer is its shape. A small rock that you hold in your hand can have an irregular shape, as its self-gravitational pull is not large enough to deform it into a sphere. However, if the rock were the size of Pluto, then gravity would indeed dominate, and it would be impossible to structure the planetoid so that it had anything other than a spherical profile. Consequently, cubical planets such as the home world of Bizarro must be very small. In fact, the average distance from the center of the Bizarro planet to one of its faces can be no longer than 300 miles, if it is to avoid deforming into a sphere. However, such a small cubical planet would not have sufficient gravity to hold an atmosphere on its surface, and it would be an airless rock. Since we have frequently seen that the sky on the Bizarro world is blue like our own (and shouldn't it be some other color if it is to hold true to the Bizarro concept?) this would imply that there is indeed air on this cubical planet. We must therefore conclude that a Bizarro planet is not physically possible, no matter how many times we may feel in the course of a day that we have been somehow instantly transported to such a world.

Back to normal spherical planets like Krypton. If the acceleration due to gravity on Krypton g_K is 15 times larger than the acceleration due to gravity on Earth g_E, then the ratio of these accelerations is $g_K/g_E = 15$. We have just shown that the acceleration due to gravity of a planet is $g = Gm/d^2$. The distance d that we'll use is the Radius R of the planet. The mass of a planet (or of anything for that matter) can be written as the product of its density (the Greek letter ρ is traditionally used to represent density) and its volume, which in this case is the volume of a sphere (since planets are round). Since the gravitational constant G must be the same on Krypton as on Earth, the ratio g_K/g_E is given by the following simple expression:

$$g_K / g_E = [\rho_K R_K] / [\rho_E R_E] = 15$$

where ρ_K and R_K represent the density and radius of Krypton and ρ_E and R_E stand for the Earth's density and radius, respectively. When comparing the acceleration due to gravity on Krypton to that on Earth, all we need to know is the product of the density and radius of each planet. If Krypton is the same size as Earth, then it must be 15 times denser, or if it has the same density, then it will be 15 times larger.

Now if, as we have argued at the start of this book, the essence of physics is asking the right questions, then it is as true in physics as in life that every answer one obtains leads to more questions. We have determined that in order to account for Superman's ability to leap 660 feet (the height of a tall building) in a single bound on Earth, the product of the density and radius of his home world of Krypton must have been 15 times greater than that of Earth. We next ask whether it is possible that the size of Krypton is equal to that of Earth ($R_K = R_E$) so that all of the excess gravity of Krypton can be attributed to its being denser than Earth ($\rho_K/\rho_E = 15$ times denser, to be precise). It turns out that if we assume that the laws of physics are the same on Krypton as on Earth (and if we give up on that, then the game is over before we begin and we may as well quit now!), then it is extremely unlikely that Krypton is 15 times denser than Earth.

We have just made use of the fact that mass is the density multiplied by volume, which is just another way of saying that density is the mass per unit volume of an object. Now, to understand what limits this density, and why we can't easily make the density of Krypton 15 times greater than Earth's, we have to take a quick trip down to the atomic level. Both the total mass of an object and how much volume it takes up are governed by its atoms. The mass of an object is a function of how many atoms it contains. Atoms are composed of protons and neutrons inside a small nucleus, surrounded by lighter electrons. The number of positively charged protons in an atom is balanced by an equal number of negatively charged electrons. Electrons are very light compared to protons or neutrons, which are electrically uncharged particles that weigh slightly more than protons and reside in a nucleus. (We'll discuss what the neutrons are doing in the nucleus in chapter 15.) Nearly all the mass of an atom is determined by the protons and neutrons in its nucleus, because electrons are nearly 2,000 times lighter than protons.

The size of an atom, on the other hand, is determined by the electrons or, more specifically, their quantum mechanical orbits. The diameter of a nucleus is about one trillionth of a centimeter, while the radius of an atom is calculated by how far from the nucleus one is likely to find an electron, and is about ten thousand times bigger than the nucleus. If the nucleus of an atom were the size of a child's marble (diameter of 1 cm) and placed in the end zone of a football field, the radius of the electron's orbit would extend to the opposite end zone, 100 yards away. The spacing between atoms in a solid is governed essentially by the size of the atoms themselves (you can't normally pack them any closer than their size).

Thus, if quantum mechanics is the same on Krypton as on Earth, the space taken up by a given number of atoms in a rock (for example) will not depend significantly upon which planet the rock resides on. The rock will weigh more on a planet with a larger gravity, but the number of atoms it contains—as well as the spacing between the atoms, both of which determine its density—will be independent of which planet the rock finds itself on. Because the number of atoms also determines the mass of the rock, it follows that the density of any given object will be the same, regardless of the planet of origin. Most solid objects have roughly the same density, at least within a factor of ten. For example, the density of water is 1 gram/cm^3 while the density of lead is 11 gram/cm^3 (a gram is one thousandth of a kilogram). In other words, a cube that measures 1 cm on each side would have a mass of 1 gram if composed of water and 11 grams if composed of lead. This higher density of lead is due almost entirely to the fact that a lead atom is ten times more massive than a water molecule. While there is a lot of water on the surface of the Earth, there's even more solid rock within the planet, so that Earth's average density is 5 gram/cm^3. In fact, Earth is the densest planet in our solar system, with Mercury and Venus close behind. Even if Krypton were solid uranium, it would have an average density of 19 gram/cm^3, which is less than four times as large as Earth's. In order for Krypton to have a gravity 15 times greater than Earth's due to a larger density alone, it would have to have a density of 75 gram/cm^3, and no normal matter is this dense.

If the density of the planet Krypton is the same as that of the planet Earth, then in order to account for the heavier gravity on Krypton, its radius must have been 15 times larger than Earth's. However, it turns out that this is no easier to accomplish than adjusting the density. While planets in our own solar system come in all sizes—from Pluto with a radius one fifth as large as Earth's, making it just barely bigger than some moons, to Jupiter with a radius of more than eleven times Earth's—the geology of the planet is a sensitive function of its size. Planets bigger than Uranus with a radius four times larger than Earth's include Neptune, Saturn, and Jupiter. These planets are gas giants, lacking a solid mantle upon which buildings and cities may be constructed, let alone supporting humanoid life. In fact, if Jupiter were ten times larger, it would be the size of our own sun. In this case, the gravitational pressure at Jupiter's core would initiate nuclear fusion, the process that causes our sun to shine. So, if Jupiter were just a bit larger, it would no longer be a giant planet but rather a small star. Big planets are gaseous because if you're going to build a very big planet, you are going to need a lot of atoms, and nearly all of the raw materials available are either hydrogen or helium gas. To be precise, 73 percent of the elemental mass in the universe is hydrogen and 25 percent is helium. Everything else that you would use to make a solid planet—such as carbon, silicon, copper, nitrogen, and so on—comprises only 2 percent of the elemental mass in the known universe. So big planets are almost always gas giants, which tend to have orbits far from a star, where the weaker solar radiation cannot boil away the gaseous surfaces they have accreted. The concentration of heavier elements with which solid planets can form is much lower, so they will tend to be smaller and closer to a star. If these inner solid planets got too large, the gravitational tidal forces from their sun would quickly tear them apart. Krypton's advanced civilization, with scientists capable of constructing a rocket ship barely large enough to hold a single infant, couldn't arise on a gas giant with a radius 15 times that of Earth's.

So, is that it? Is the story of Superman and Krypton, with an Earth-like surface and a gravity 15 times that of Earth, totally bogus? Not necessarily. Remember that earlier it was stressed that no *normal* matter could be 15 times denser than matter on Earth. However, astronomers have discovered *exotic* matter, with exceedingly high densities, that is the remnants of supernova explosions. As mentioned, when the size of a gaseous planet exceeds a certain threshold, the gravitational compression at its center is so large that the nuclei of different atoms literally fuse together, creating larger nuclei and releasing excess energy in the process. The source of this energy is expressed in Einstein's famous equation, $E = mc^2$ or Energy E is equivalent to mass m multiplied by the speed of light c squared. The mass of the fused-product nucleus is actually a tiny bit smaller than that of the two initial separate nuclei. The small difference in mass, when multiplied by the speed of light squared (a very big number) yields a large amount of energy. This energy radiates outward from the star's center, producing an outward flow that balances the inward attractive gravitational force, keeping the radius of the star stable. When all the hydrogen nuclei have been fused into helium nuclei, some of the helium nuclei are in turn fused into carbon nuclei, some of which in turn are compressed to form nitrogen, oxygen, and all of the heavier elements, up to iron. The fusion process speeds up as the star generates heavier and heavier nuclei, so that all of its iron and nickel are created within the last week of the star's life. As heavier and heavier nuclei are combined, the process becomes less and less efficient, so that the energy released when iron nuclei fuse is insufficient to stably counteract the inward gravitational pressure. At this point gravity wins out, rapidly compressing the material into a much smaller volume. In this brief moment the pressure at the center of the star is so high that one last gasp of fusion occurs, and heavier elements all the way up to uranium are generated, with a concurrent tremendous release of energy. This last stage in the life of a star is termed the "supernova" phase. With this final blast of energy, the elements that had been synthesized within the star are flung out into space, where gravity may eventually pull them together into clumps that can form planets or other stars. Every single atom in your body, in the chair in which you are sitting, or the paper and ink in *Action* Comics # 1, was synthesized within a star that died and subsequently expelled its contents. Thus, we are all composed of stardust or, if you're feeling a tad more cynical, solar excrement.

For really big stars the gravitational pressure at the center of the star is so great that even after the supernova phase, there remains a large remnant core for which gravity compresses the protons and electrons into neutrons, which are squeezed until they touch and become a solid composed of nuclear matter. The remnants of such massive stars are termed "neutron stars," and their density is exceeded only by that of black holes (left over from the death of even bigger stars, whose gravitational attraction is so high that not even light can escape its pull). Compared to the density of lead (11 grams per cubic centimeter), the density of neutron star material is one hundred thousand billion grams per cubic centimeter. That is, a teaspoon of neutron star material on Earth would weigh more than 100 million tons. This is just the stuff to boost the gravity of Krypton.

If a planet the size of Earth had a small volume of neutron star material within its core, the additional mass would dramatically increase the gravity on the surface of the planet. In fact, it would only take a sphere of neutron star material with a radius of 600 meters (about the length of six football fields) at the center of a planet the size of Earth to create an acceleration due to gravity on its surface of 150 meters/sec^2, whereas the acceleration due to gravity on Earth is 10 meters/sec^2. So, in order for Krypton's gravity to have been 15 times greater than on Earth, it must have had a core of neutron star matter at its center.

And thus we see why Krypton exploded! For such a super-dense core would produce enormous strains on the surface of the planet, making a stable distribution of matter tenuous at best. At some point in the planet's history, volcanic activity and plate tectonics would result in massive upheavals. Such preshock earthquakes would warn scientists that now would be a good time to put their infant children into a rocket and send them to some other distant planet, preferably one without a neutron star core.

Let us pause to admire the scientific insight of Jerry Siegel and Joe Shuster. These teenagers from Cleveland, Ohio, either had an understanding of astrophysics and quantum mechanics that exceeded that of many contemporary physics professors in 1938, or they were very lucky guessers. Only eight years earlier Subrahmanyan Chadrasekhar had calculated the minimum radius of a star for which its post-supernova remnants would form a white dwarf, a calculation for which he would be honored with a Nobel Prize in 1984. Perhaps if Sheldon Mayer at National Publications had not taken a chance on their Superman strip, Siegel and Shuster might have considered publication in a scientific journal such as the *Physical Review*, and the history of both science and comic books would be very different today.

THE DAY GWEN STACY DIED–

IMPULSE AND MOMENTUM

IF A SENATE SUBCOMMITTEE HEARING marked the beginning of the end of the Golden Age of comics, the death blow of the Silver Age was self-inflicted. Viewed from today's perspective, comics from the Silver Age (from the late 1950s to 1960s) seem suffused with an optimistic outlook and a sunny disposition that borders on the Pollyanna-esque. The Golden Age characters reinvented by Julius Schwartz and colleagues at DC comics in the late fifties and early sixties, such as the Flash, Green Lantern, or Green Arrow (an amalgam of Batman and Robin Hood, with a quiver full of gadget arrows such as a "boxing glove arrow" or a "handcuff arrow" whose successful application violated several fundamental principles of aerodynamics) carried on the positive outlooks and righteousness of their Golden Age antecedents, and their plot-driven twelve- or twenty-two-page-long stories did not leave much room for character development. A typical Silver Age hero in a DC comic book would gain superpowers through some implausible mechanism and then decide, as a matter of course, to use said powers to fight crime and better humanity (after first donning, of course, a colorful costume), never questioning this career choice.

The situation was very different with the superheroes populating DC's main comic-book competitor, Marvel Comics, whose characters such as the Hulk and the X-Men lamented that if they didn't have bad luck, they'd have no luck at all. In 1961, the Marvel Comics (né Timely) Company was on the verge of going out of business. From its Golden Age peak when it had published the

Human Torch, the Sub-Mariner, and Captain America comics, it had fallen to the point where it was barely getting by putting out monster comics, westerns, funny animal stories, and young romance stories. This all changed with a golf game between Jack Liebowitz, the head of DC Comics and Martin Goodman, Marvel Comics' publisher. Liebowitz boasted of the success DC was having with a particular title, the *Justice League of America*, that featured a team of superheroes including Wonder Woman, the Flash, Green Lantern, Aquaman, the Martian Manhunter, and others banding together to fight the supervillain *du mois*. Upon returning to the office, Goodman instructed his editor (and his nephew-in-law, the last remaining full-time employee) Stan Lee to come up with a comic book featuring a team of superheroes. Marvel was not publishing superhero comics at the time; consequently Lee could not assemble a team book by incorporating characters from other comics as DC had done. Instead he created a new superhero team from whole cloth. The resulting title, the *Fantastic Four*, written by Lee and with art by Jack Kirby, became a sales success, and led to Marvel Comics' reversal of fortune.*

Lee's and Kirby's unique contribution was to add character development and distinctive personalities to their comic-book stories. As a way of distinguishing themselves from the heroes in DC comics, the superheroes in Marvel's stories did not see their superpowers as a blessing, but frequently bemoaned their fate. When cosmic radiation turned Ben Grimm into the large, orange, rock-complexioned Thing in *Fantastic Four* # 1, he did not revel in his newfound superhuman strength but cursed the fact that he had become a walking brick patio, wanting nothing more than to be restored to his human form. But no character in the Marvel universe complained more about his lot in life than Spider-Man.

In *Amazing Fantasy* # 15 written by Lee and drawn by Steve Ditko in 1962, young Peter Parker was a thin, nerdy high school student who suffered endless teasing by the popular jocks at his school. Parker was an orphan, living with his overprotective,

---

* Those who were involved in publishing both DC and Marvel comics at the time deny that such a golf game ever took place. Nevertheless, because this story is considered the fountainhead of Marvel Comics by so many fans, it has become the accepted legend, regardless of whether it has any factual basis.

elderly relatives, Aunt May and Uncle Ben. Excluded from joining the popular students in an after-school activity, Peter indulged his interest in science by attending a physics lab demonstration on radioactivity. As happened with alarming frequency in Silver Age Marvel comics, an accident with radioactivity resulted in the bestowal of superpowers. In this case, a spider was inadvertently irradiated during the demonstration and bit Peter Parker before expiring, leaving him with radioactive spider blood.

Parker found that he had acquired from this spider bite various arachnid attributes, including heightened agility and the ability to adhere to walls. Because a spider can lift several times its own body weight, Peter could now lift several times his own weight, described in the comics as a "proportionate" increase in strength. Peter also gained a "sixth sense" that alerted him to potential dangers, a Spider-Sense, if you will. One can only guess that Stan Lee, stymied in his attempts to kill real spiders in his bathroom, concluded that spiders used ESP to avoid being squished. Presumably we can thank the protective Comics Code Authority for the fact that Peter did not also gain a spider's ability to spew organic webbing from his anus, but instead used his knowledge of chemistry and mechanics to construct technological web shooters that he wore on his wrists.*

After a lifetime of ridicule and abuse at the hands of his peers, Peter initially sees his newfound powers as a venue to fame and fortune. After testing his skills in professional wrestling, he creates a colorful blue-and-red costume and mask in order to enter show business. Feeling empowered on the eve of his television debut, he arrogantly refuses to help a security guard stop a fleeing robber, though it would have been easy to do so. However, upon returning home, he learns that gentle Uncle Ben has been slain by an intruder. Capturing the killer using his new powers, Peter discovers to his horror that this was the same robber he could have stopped earlier that day. Belatedly realizing, as his uncle

---

* In the 2002 motion picture *Spider-Man*, the genetically engineered super-spider bite also gave Peter the ability to shoot organic webbing from ducts in his wrists. This freed the filmmakers from having to explain why teenager Peter Parker was able to invent and manufacture a revolutionary adhesive webbing, yet persistently remained in debt (a paradox that never concerned the average Silver Age comic-book reader).

had presciently taught him earlier, "that with great power there must also come great responsibility," Peter Parker dedicates himself to fighting crime and righting wrongs as the amazing Spider-Man.

Not that he didn't continue to complain about his life at least three times per issue. One of the novelties that Lee and Ditko introduced in the Spider-Man comic book was a host of real-life concerns and difficulties that bedeviled Spider-Man nearly as much as his colorful rogue's gallery of supervillains. Peter Parker would contend with seething high school romances and jealousies, money problems, anxiety over his aged aunt's health, allergy attacks, even a sprained arm (he spent issues # 44–46 of the *Amazing Spider-Man* with his arm in a sling), all while trying to keep the Vulture, the Sandman, Doctor Octopus, and the Green Goblin at bay. But the greatest intrusion of reality, which would signal the end of the innocent Silver Age, would come in 1973 in *Amazing Spider-Man* # 121 with the death of Peter Parker's girlfriend, Gwen Stacy. A death that was demanded, as we will now show, not by the writers and editors or by the readers, but rather by Newton's laws of motion.

The Green Goblin had first appeared in *Amazing Spider-Man* # 14 as a mysterious crime over-boss, and grew into one of Spidey's most dangerous foes. In addition to enhanced strength and an array of technological weapons, such as a rocket-propelled glider and pumpkin bombs, the Green Goblin managed to unmask Spider-Man and learn his secret identity in the classic *Amazing Spider-Man* # 39. Knowing that Peter Parker was really Spider-Man gave the Goblin a distinct advantage in his battles. In *Amazing Spider-Man* # 121, the Goblin kidnaps Parker's girlfriend, Gwen Stacy, and brings her to the top of the George Washington Bridge, using her as bait to lure Spider-Man into battle. At one point in their fight, the Goblin knocks Gwen from the tower, causing her to fall to her apparent doom (see figs. 6 and 7).

At the last possible instant, Spider-Man manages to catch Gwen in his webbing, narrowly preventing her from plummeting into the river below. And yet, upon reeling her back up to the top of the bridge, Spider-Man is shocked to discover that Gwen is in fact dead, despite his last-second catch. "She was dead before your webbing reached her!" the Goblin taunts. "A fall from that height would kill anyone—before they struck the ground!" Apparently the Green Goblin, creator of such advanced

Fig. 6. *Gwen Stacy's fatal plunge off the top of the George Washington Bridge, as told in* Amazing Spider-Man # 121. *Note the "SNAP" near her neck in the second to last panel.*

© 1973 Marvel Comics

Fig. 7. Continuation of Gwen Stacy's death scene, where Spider-Man receives a harsh physics lesson, and the Green Goblin's scientific "genius" is called into question.

technology as the Goblin-Glider and Pumpkin Bombs, suffers from a basic misunderstanding of the principle of conservation of momentum.

Of course, if it were true that it was "the fall" that killed poor Gwen, then the implication for the fate of all skydivers and paratroopers would suggest a massive conspiracy of silence on the part of the aviation industry. Nevertheless, comic-book fans have long argued over whether it was indeed the fall or the webbing that killed Gwen Stacy. This question was listed as one of the great comic-book controversies (comparable to whether the Hulk is stronger than Superman) in the January 2000 issue of *Wizard* magazine. We now turn to physics to definitively resolve the question of the true cause of the death of Gwen Stacy.

The central question we here pose is: How large is the force supplied by Spider-Man's webbing when stopping the falling Gwen Stacy?

PHYSICS AND THE FINAL FATE OF GWEN STACY

To determine the forces that acted upon Gwen Stacy, we first need to know how fast she was falling when the webbing stopped her. In our previous discussion of the velocity required for Superman to leap a tall building in a single bound, we calculated that the necessary initial velocity v was related to the final height h (where his speed is zero) by the expression $v^2 = 2gh$ where g is the acceleration due to gravity. The process of falling from a height h with initial velocity $v = 0$, speeding up due to the constant attractive force of gravity, is the mirror image of the leaping processes that got him to the height h in the first place.

This is why most cities outlaw the shooting of firearms, even as part of a New Year's Eve celebration. A bullet leaving a gun with a velocity of 1,500 feet/sec slows down due to gravity, and then speeds up as it falls, until it strikes the ground. The final velocity is less than the initial speed, due to some energy lost to air resistance (on the way up, this loss of energy by air drag decreases slightly the final height to which the bullet climbs). But what goes up must come down, and it will land with close to the same speed it had when taking off.

The upshot is that we can employ the expression $v^2 = 2gh$ to calculate Gwen Stacy's speed right before she is caught in Spider-Man's webbing. Assuming that Spidey's webbing catches her after she has fallen approximately 300 feet, Gwen's velocity turns out to be nearly 95 mph. Again, air resistance will slow her down somewhat, but as indicated in fig. 6, she is falling in a fairly streamlined trajectory. As we are about to discuss, the danger for Gwen is not the speed but the sudden stopping that the river would provide.

In order to change Gwen Stacy's motion from 95 mph to zero mph, an external force is required, supplied by Spider-Man's webbing. The larger the force, the greater will be the change in Gwen's velocity, or rather, her deceleration. To calculate how large a force is needed in order to bring Gwen to rest before she strikes the water, we once again turn to Newton's second law, **F = ma**. Recall that the acceleration is the change in velocity divided by the time during which the speed changes. Multiplying both sides of the expression **F = ma** by the time over which the speed decreases, we can rewrite Newton's second law as

(FORCE) X (TIME) = (MASS) X (CHANGE IN SPEED)

The momentum of an object is defined as the product of its mass and its speed (the right-hand side of the previous equation). The product of **(Force) × (time)** on the left-hand side of this equation is called the **Impulse**. The previous equation, therefore, tells us that in order to change the momentum of a moving object, an external force **F** must be applied for a given time. The larger the interval of time, the smaller the force needed to achieve the same change in momentum.

This is the principle behind the air bags in your automobile. As your car travels down the highway at a speed of, say, 60 mph, you as the driver are obviously also moving at this same speed. When your car strikes an obstacle and stops, you continue to move forward at 60 mph, for an object in motion will remain in motion unless acted upon by an external force (coming up in an instant). In the pre–seat belt and air bag days, the steering column typically supplied this external force. The time your head spent in contact with the steering wheel was brief, so consequently the force needed to bring your head to rest was large. By rapidly inflating an air bag, which is designed to deform under pressure, the time your head remains in contact with the inflated air bag increases, compared to the steering wheel, so the force needed to bring your head to rest decreases. Distributing the force over the larger surface area of the air bag also helps to reduce injuries in a sudden stop. This force is still large enough to often knock the driver unconscious, but the important point is that it is no longer lethal. The product of force and time must always be the same, as the net result is the same— namely, the initial velocity of 60 mph changing to a final velocity of zero. This is also the physical justification for a boxer rolling with a punch, increasing the time of contact between his face and his opponent's fist, so that the force his face must supply to stop the fist is lessened.

Now, Spider-Man's webbing does have an elastic quality, which is a good thing for Gwen Stacy, but the time that is available to slow her descent is short, which is an awful thing. The shorter the time, the greater the force must be to achieve a given change in momentum. For Gwen, her change in speed is 95 mph − 0 mph = 95 mph, and assuming she weighs 110 pounds, her mass in the metric system is 50 kilograms. If the webbing brings her to rest in only about 0.5 seconds, then the force applied by the webbing to break her fall is 970 pounds. Hence, the webbing applies a force nearly ten times larger than Gwen's weight of 110 pounds. Recalling that an object's weight is simply $\mathbf{W = mg}$, where $\mathbf{g}$ is the acceleration due to gravity, we can say that the webbing applies a force equivalent to 10 $\mathbf{g}$'s in a time span of 0.5 seconds. As indicated in fig. 6, when the webbing brings Gwen to a halt, a simple sound effect drawn near her neck (the "SNAP!" heard round the comic-book world) indicates the probable outcome of such a large force applied in such a short period of time. In contrast, bungee jumpers allow a sufficient distance to enable the stretching cord to extend for many seconds, in order to keep the braking force below a fatal threshold.

Traveling at such a speed, coming to rest in such a short time interval, there is no real difference between hitting the webbing and hitting the water. However, there have been recorded cases of people surviving forces greater than that experienced by Gwen Stacy. Col. John Stapp rode an experimental rocket sled in 1954 and was subjected to a force of 40 $\mathbf{g}$'s during deceleration, yet lived to describe the experience as comparable to "dental extraction without anesthetic." Of course, Colonel Stapp was securely strapped into and supported by the sled in a reinforced position. More typically, suicide victims who jump from bridges die not from drowning, but rather from broken necks. Hitting a body of water at such a speed has the same effect as hitting solid ground, as the fluid's resistance to displacement increases the faster you try to move through it (we'll discuss this further in chapter 5 when considering the Flash). Tragically for Gwen Stacy, and for Spider-Man, this is another example of when comic books got their physics right, and we readers were not required to suspend our disbelief, no matter how much we may have wanted to.

Spider-Man seems to have learned this physics lesson concerning Impulse and change in momentum. A story in *Spider-Man Unlimited* # 2, entitled appropriately enough "Tests," finds the wall-crawler adhering to the top of a skyscraper when an unfortunate window washer plummets past him. Launching himself after the falling worker, Spider-Man must solve a real-life physics problem under more pressure than in a typical final exam. As he closes the gap between the worker (due to the fact that Spider-Man pushed off from the building with a larger initial velocity than did the window washer), Spider-Man considers: "OK, I have to do this right. Can't snag him with a web-line, or the whiplash will get him." As shown in fig. 8, Spider-Man recognizes that his best solution is to match his speed to that of the worker and then grab hold of him when they are barely moving relative to each other. (I'm not sure how Spider-Man slows himself down to match the worker's velocity—perhaps by dragging his feet against the side of the building?) Then Spidey shoots out a web-line, where his arm, endowed with spider-strength, is able to withstand the large Impulse associated with their upcoming change of momentum.

This solution was also employed in the 2002 motion picture *Spider-Man*. When the Green Goblin drops Mary Jane Watson from a tower on the Queensboro Bridge, in a clear homage to the storyline from *Amazing Spider-Man* # 121, Spider-Man this time does not stop her rapid descent with his webbing. Rather he dives after her, and only after catching her does he employ his webbing to swing them to (relative) safety, using the same procedure as in fig. 8. One hallmark of a hero, it appears, is the ability to learn from experience.

While certainly no hero, the above arguments have also made an impression on the Green Goblin. As mentioned earlier, the January 2000 issue of *Wizard* magazine described the controversy surrounding the death of Gwen Stacy as a classic open question in comic-book fandom. This prompted my letter to the editor of *Wizard*, published a few months later, which summarized the above physics discussion. Two years later, the August 2002 issue of *Peter Parker: Spider-Man* # 45 (written by Paul Jenkins and drawn by Humberto Ramos) featured a story line in which the Green Goblin demonstrated that he had also finally learned this physics lesson. In this issue the Goblin had sent a videotape of Gwen Stacy's death to the news media in order to psychologically torment

Fig. 8. *Scene from the story "Tests" in* Spider-Man Unlimited # 2 (May 2004), *in which the caption boxes reveal Spider-Man's thought process as he faces a practical application of Newton's second law of motion.*

Spider-Man. Portraying himself as the reluctant hero of this tragedy, the Goblin narrates in the tape:

> "Realizing the girl had fallen, I naturally made a course correction on my glider in an attempt to save her. I began an immediate descent. But before I had a chance to reach her, Spider-Man did something incredibly stupid: despite the speed of her fall, he chose to catch her in that rubber webbing of his. In the next instant, her neck was snapped like a rotten twig."

It may have taken the Goblin nearly thirty years, but apparently he at last understands that it wasn't "the fall" that killed Gwen Stacy, but the sudden stopping. If a twisted, evil maniac like the Green Goblin can learn his physics, then there is hope for us all.

4

CAN HE SWING FROM A THREAD?—
CENTRIPETAL ACCELERATION

I WOULD LIKE TO MAKE one more point about forces, in particular as they relate to Spidey's web-swinging abilities. Practically every issue of the *Amazing Spider-Man* features scenes of him using his webbing to swing from building to building through the canyons of New York City. But is this realistic? Specifically, is Spider-Man's webbing strong enough to support his own weight, as well as the weight of any falling criminal, victim, or innocent bystander whom he happens to catch mid-flight as he swings in his parabolic trajectory? As Spider-Man swings in an arc, there is an extra force in addition to his weight that the webbing must supply. Let's now consider why.

Remember that Newton's second law of motion, $\mathbf{F} = \mathbf{ma}$, told us that a force is needed to change an object's motion. A change in motion, or acceleration, refers to either a change in magnitude (either speeding up or slowing down) or to a change in direction. If no force acts on the object, then it persists in "uniform motion," which means constant motion in a straight line. Any change in motion, whether in magnitude or direction, can only come about if a force acts on the object. When an automobile negotiates a hairpin turn, an external force (friction between the tires and the road) changes the auto's direction, even if the speed remains unchanged.

In order to change the direction of motion, an external force is needed—and the corollary of this is that a force can only produce an acceleration (a change in the motion) in the direction that it acts. For example, gravity pulls an object toward the ground, regardless of its initial motion. More importantly, gravity can *only* pull an object toward the ground, because that is the only direction it acts. If the Golden Age, pre-flying Superman runs off the edge of a cliff with a steady horizontal speed, he will start falling due to gravity. Since gravity does not act in the horizontal direction, his horizontal speed will not change as he falls! No force, no change, after all. His vertical velocity does increase, becoming greater the longer he falls, just as in the case of Gwen Stacy considered a moment ago, because there *is* a force acting on him in the vertical direction. The net effect of his constant horizontal speed plus an ever-increasing vertical speed is a parabolic trajectory that becomes steeper the longer he plummets. Put another way, a 90 mph fastball, thrown without spin perfectly parallel to the ground, falls to the ground at the exact same rate as a ball simply dropped out of the pitcher's hand at the same moment. Both balls would strike the ground at the same instant (if released from the same height), because the only force changing their motion is gravity, in the vertical direction. Any change in either the direction or magnitude of an object's motion can only arise from an external force acting in the direction of the change.

As Spider-Man swings from building to building, his trajectory is a semicircular arc rather than a straight line. Therefore, even if the magnitude of his speed does not change during his swing, his direction of motion is continually being altered, which can only be accomplished by an external force. It should be obvious that this force comes from the tension in the webbing. The webbing therefore has to do double duty and supply *two* forces: (1) supporting Spider-Man's weight, which it would have to maintain even if he were simply hanging from the vertical line, and (2) a second force to divert him in a circular trajectory. If the webbing line were to snap in mid-swing, then the only external force acting on Spider-Man is gravity, and his motion at this point would not differ at all from that of a ball tossed with the same velocity Spider-Man possessed at the instant the webbing broke.

The acceleration that this additional force in the webbing provides as Spidey swings in a circular arc is identical to the acceleration experienced by the moon as it makes its circular orbit about the Earth. In one case the force arises from the tension in the webbing, while in the other it is Newton's force of gravitational attraction, but for both it changes straight-line motion into circular motion. Gravity is the moon's "webbing," causing its direction to change.

If the tension in the webbing or gravity were suddenly to disappear, both Spidey and the moon would depart from their circular trajectories, and continue moving with the velocity they had at the moment the external forces were removed. With a little bit of geometry or calculus one can show that the acceleration $\mathbf{a}$ of an object being constantly deflected onto a circular orbit with a velocity $\mathbf{v}$ is $\mathbf{a} = (\mathbf{v} \times \mathbf{v})/R = v^2/R$, where R is the radius of the circle.

Spider-Man's webbing has to supply a force $\mathbf{mg}$, in order to support his weight, and an additional force $\mathbf{mv^2/R}$ in order to change his direction as he swings. The faster he swings (the larger his velocity $\mathbf{v}$) or the tighter his arc (that is, the smaller the radius $\mathbf{R}$), the greater will be this centripetal acceleration v^2/R. When Spidey swings from a web strand 200 feet long, at a speed of 50 mph, the centripetal acceleration is 27 feet/sec², in addition to the acceleration due to gravity of 32 feet/sec². If Spider-Man's mass in the metric system is approximately 73 kilograms, then his weight $\mathbf{mg}$ is 160 pounds, and the additional force the webbing must supply just to change his trajectory from straight-line motion into a circular arc is roughly 135 pounds. The total tension in the webbing is nearly 300 pounds, and will be more if Spidey is carrying someone as he swings.

Three hundred pounds or greater of tension may seem to be more than a thin strand of fiber can withstand, but if Spider-Man's webbing is anything like real spider silk, he has nothing to worry about. Dragline silk webbing, which spiders use for their webs and while rapidly fleeing predatory birds, is actually five times stronger per pound than steel cable and more elastic than nylon. The webbing's properties result from thousands of rigid filaments only a few billionths of a meter wide (providing great redundancy so that no one filament is crucial for the integrity of the webbing), interspersed with fluid-filled channels that distribute the tensile force along the length of the webbing. Spider-Man is able to alter the material properties of his webbing by adjusting its chemical composition as it jets from his web-shooters. Similarly, real spiders can control the tensile strength of their webs by varying the relative concentration of crystallizing and noncrystallizing proteins.

There is considerable interest in commercial applications of webbing, which would require large quantities of spider silk. As it is not practical to harvest spiders for their silk (they are too territorial to farm in a conventional manner), recent genetic engineering experiments have inserted a spider's web-making genes into goats, so that the goats' milk will contain webbing that can be more easily sieved and acquired. While the development of web-producing goats has hit some snags,* other scientists have reported preliminary success with infecting spider cells in the laboratory with a genetically engineered virus that induces the cell to directly manufacture the proteins found in spiderwebs. The silk-producing gene from spiders has also been successfully introduced into E. coli and plant cells. Such research could have far-reaching practical applications. As Jim Robbins discussed in his article "Second Nature" in the July 2002 issue of *Smithsonian*: "In theory, a braided spider silk rope the diameter of a pencil could stop a fighter jet landing on an aircraft carrier. The combination of strength and elasticity allows it to withstand an impact five times more powerful than Kevlar, the synthetic fiber used in bulletproof vests."

The high tensile strength of real spider silk enables it to support a weight of more than 20,000 pounds per square centimeter. That is, if the cross section of the webbing was a circle with a diameter of 1 cm (a little bit less than a half inch), then the webbing could hold a weight of 8 tons before breaking. Even a webbing strand with a diameter of only a quarter inch could support more than 6,000 pounds safely, well below the 300 pounds of weight and centripetal force we estimated earlier. Unless Spider-Man is trying to carry both the Hulk and the Blob simultaneously, his webbing should be more than able to do the job.

Therefore, according to Newton's laws of motion, it is entirely plausible that Spider-Man can swing from building to building, stop a runaway elevated train (as in the 2004 film *Spider-Man 2*), or weave a bulletproof shield out of very narrow lines of webbing. So, to answer the question posed in the title of this chapter: Simply take a look overhead!

---

* Sorry.

5

FLASH FACTS—
FRICTION, DRAG, AND SOUND

IT WAS A DARK AND STORMY night in Central City as police scientist Barry Allen locked up for the night. Pausing by the chemical storeroom, he marveled at the large collection of chemicals that the CCPD possessed. Despite his scientific training, Allen was standing near an open window during the gathering storm and bore the full brunt of a lightning strike that entered the room. The lightning bolt shattered the chemical containers, dousing him while the electrical current passed through his body.

But the simultaneous exposure to lethal voltages and hazardous chemicals somehow only dazed Allen and knocked him off his feet. Later that evening he was surprised to discover that he could easily outrace a departing taxicab and catch and restore a spilled plate of food in a diner in the blink of an eye. Realizing that the lab accident had somehow endowed him with super-speed, he adopted a simple yet elegant red-and-yellow costume and used his newfound powers to fight crime as the Flash.*

There is a broad range of physical phenomena associated with speed, and John Broome, Robert Kanigher, and Gardner Fox, the main writers of the early Silver Age *Flash* comics, addressed many

---

* A freak electrochemical accident of this nature would not reoccur until *Flash* comics # 110, when *another* lightning bolt splashed young Wally West with identical chemicals, endowing him with super-speed as well. Wally then began his career as a junior crimefighter, under the imaginative name Kid Flash.

of them. Thanks to his ability to run very fast, the Flash was frequently depicted running up the sides of buildings or across the ocean's surface; he would catch bullets shot at him, and drag people behind him in his wake. Are any of these feats consistent with the laws of physics? It turns out that all of them are, granting of course the one-time "miracle exception" of the Flash's super-speed in the first place.

* * *

In his very first Silver Age appearance, "The Mystery of the Human Thunderbolt" in *Showcase* # 4, the Flash ran up the side of an office building, because with his "great speed he is able to overcome gravity." Earlier we explored the simple relationship between someone's initial vertical velocity and the final height he can leap. As the person rises, he slows down due to gravity, until at a height **h** the final speed is zero. We calculated in chapter 1 that for Superman to leap a height of 660 feet, equivalent to a thirty- to forty-story building, his initial liftoff velocity needs to be at least 140 mph. But the Flash can run much, much faster than this, and he should therefore be able to reach the top of a forty-story building with velocity to spare. So, as he approaches the side of a building, as long as he has a speed greater than the minimum $v^2 = (2gh)$, he should be able to leap up its side without violating any laws of physics (aside from the fact that he *is* running several hundred miles per hour, that is). In contrast, the fastest that a non-superpowered human can run is on the order of 15 mph (though faster sprints are possible)—which would enable him or her to run up the side of a small tool shed.

The catch is not whether the Flash is able to move fast enough to *leap* a vertical height **h**, but whether he can maintain traction to actually *run* up the vertical side of the building. Some interesting physics underlies the simple act of walking, related to Newton's third law, which states that forces come in pairs. When you run or walk, a force must be applied horizontal to the ground by your feet, opposite to the direction you wish to move. The ground exerts an equal and opposite force back on your feet, parallel to the ground's surface, that counters the back-directed force exerted by your shoes. The origin of this parallel force is friction. Imagine trying to walk across a floor covered with a uniform layer of motor oil, and you will realize how crucial friction is to a process as simple as walking. Without friction between his boots and the ground, the Flash would never be able to run anywhere. Captain Cold, one of the first and most persistent supervillains that the Flash would regularly combat, possessed a "freeze ray" gun that could ice up any surface. Time and again, Captain Cold (who, incidentally, isn't really a captain) would use the simple act of creating a layer of ice directly in front of the Scarlet Speedster to immobilize him, denying him traction and rendering his super-speed useless.

No doubt due to its ubiquity and fundamental role in everyday life, the phenomenon of friction is generally taken for granted, despite its complexity. Exactly why does an object resist being dragged across another surface? While friction's basic properties were first scientifically addressed by Leonardo Da Vinci in the early 1500s and Amontons in the mid-1600s, a true understanding of the root cause of this phenomenon would not arrive until the atomic nature of matter was properly resolved in the 1920s.

There are primarily two ways in which atoms can be arranged to form a macroscopic object: (1) in a uniform, periodic, crystalline structure or (2) in a random, amorphous agglomeration. Of course, most matter lies somewhere between these two extremes, and typically there will be regions of crystalline order randomly connected, sometimes separated by amorphous sections. The net result will be that even the smoothest macroscopic surface will not be truly flat when viewed on an atomic scale. In fact, one doesn't have to go to such extremes: even on length scales of a thousandth of a millimeter—much, much bigger than an individual atom—an object's surface will more likely resemble a jagged mountain range than the stillness of a quiet lake. Consequently, when two objects are dragged past each other, regardless of the apparent smoothness of their finishes, on the atomic scale it is not unlike taking the Rocky Mountain range, turning it upside down, shoving it atop the Himalayas, and then dragging the upside-down Rockies at a steady speed across the Himalayan mountaintops. One would naturally expect enormous geological upheavals and large-scale distortions in this extreme form of plate tectonics, and the results are no less catastrophic at the atomic level. With every footstep, bonds between atoms are broken, new bonds are formed, and atom-size avalanches and atom-quakes are produced (to say nothing of the atomic Armageddon that results from tap-dancing). All of this requires a great deal of force in order to keep these atomic-scale mountain ranges sliding past and through each other. The resistance to such atomic rearrangements is called "friction," and without it, the Flash would only be running in place.

The amount of friction opposing the motion of an object along a horizontal surface is proportional to the weight of the object pressing down on the surface. The greater the weight of an object, the deeper the atomic "mountain ranges" interpenetrate, and the greater the frictional force that must be overcome to move the object. Even when a large enough force is applied, it is harder to get a big, heavy block to start moving than a smaller, lighter one. Engineering solutions for overcoming the large friction of heavy objects have been known since the time of the ancient Egyptians, who developed various ingenious schemes for moving giant limestone blocks during the pyramids' construction.

One obvious trick is to use a ramp. On a horizontal flat surface, all of a block's weight presses down perpendicular to the surface. On a sloped surface, on the other hand, the weight is still straight down, directed toward the center of the Earth (think of a plumb line held on the ramp). Only some of the weight is perpendicular to the surface of the tilted ramp, and the rest is directed down the ramp. The smaller the force pressing the atomic mountain ranges against each other, the less they will interpenetrate, and the easier it will be to move them past each other. So the frictional force, which is proportional only to the component of the weight perpendicular to the surface, is less for a block on a tilted surface compared with on a horizontal surface. No matter how rough the surface, if the ramp is tilted at too steep an angle, the friction force holding the block in place will be insufficient to counteract the downward pull of the weight down the ramp, and the block will slide down the ramp. However, as the Flash runs up the vertical side of a building, there is *no* component of his weight perpendicular to the surface upon which he is running, that is, the building's face. In principle, therefore, there should be no friction between his boots and the building's wall, and without friction he cannot run at all.

So can he in fact run up the side of a building? Technically, no. At least, not "run" as we understand the term. He can, as he leaps up the side of the building, move his feet back and forth against the building's side, which would make it appear as if he were running. In essence he is traveling a distance equivalent to the height of the building in the time between steps. Typically as the Flash runs, his foot pushes down on the ground at an angle with the road's surface, so that the force the road exerts back on him (thanks to Newton's third law) is also at an angle with the surface. The net effect is that he accelerates in both the vertical and horizontal direction. The vertical velocity gives him a bounce up off the ground, and the horizontal component propels him in the direction he is running. The greater the vertical velocity, the higher the bounce, while the larger the horizontal velocity, the farther he advances before gravity overcomes the small vertical velocity and brings his feet back to the ground, ready for another step. Very fast runners, which would certainly include the Flash, can have both feet up off the ground between steps. The faster they run, the longer their time "airborne" between steps. If the Flash bounces about 2 cm vertically with every step, then he is in the air for about one eighth of a second before gravity pulls him down for another step. But one eighth of a second is a long time for the Scarlet Speedster. If his horizontal velocity is 5,250 feet/sec or 3,600 mph, then the distance he travels between steps is more than 660 feet. This is approximately one eighth of a mile, which we used as the benchmark for the tall building that Superman leaped in chapter 1. As long as the Flash maintains at least this minimum speed, he needn't worry about losing his footing along the way, simply because he will scale the height of the building between steps.

Before he can scale a skyscraper, the Flash has to radically alter his direction from the horizontal to the vertical. As emphasized in the previous chapter, any change in the direction of motion, whether it is Spider-Man swinging on his webbing or the Flash changing his path up the side of a building, is characterized by an acceleration that requires a corresponding force. Rotating his trajectory by ninety degrees up the building's face entails a large force, provided by the friction between the Crimson Comet's boots and the ground. In addition to super-speed, the Flash's "miracle exception" must therefore also extend to his being able to generate and tolerate accelerations that few superheroes not born on Krypton could withstand.

Newton's laws of motion can also explain how the Flash is able to run along the surface of the ocean, or any body of water for that matter. Just as Gwen Stacy had to be concerned as she was about to strike the water moving at her large, final velocity, the great speed of the Flash's strides enables him to run across its surface. As one moves through any fluid, be it air, water, or motor oil, the fluid has to move out of your way. The denser the medium, the harder it is for this to be accomplished quickly. It requires more effort to walk through a swimming pool, pushing the water out of your way, than to walk through an empty pool (that is, one filled only with air) and it is harder still if the swimming pool is filled with molasses. The resistance of a fluid to flow is termed "viscosity," which typically increases the denser the medium and the faster one tries to move through the fluid.

For a dilute medium such as air, there is a lot of space between neighboring molecules. At room temperature and pressure, for example, the distance between adjacent air molecules is about ten times larger than the diameter of an oxygen or nitrogen molecule. Moreover, each air molecule at room temperature is zipping around with an average speed of approximately 1,100 feet/sec or 750 mph (which is the speed of sound in air). When we run through air, we don't build up a high-density region in front of us, because our speed is much less than the average air molecule's velocity. Think about herding cattle: If the cows are running when you try to push one into the herd, the others will just run away. If they are walking slowly, and you push at the same rate, the others don't have time to get out of the way, and they pile up into a herd. One can, of course, move faster than the speed of sound (a feat first performed by Col. Chuck Yeager in 1947), but the expended effort is large. When trying to displace a volume of air faster than the air molecules are moving, a high-density region (that is, a shock front) piles up in front of you.

In fact, in "The Challenge of the Weather Wizard" the Flash uses just such a shock front to knock out the Weather Wizard. Mark Mardon is a small-time crook who steals his deceased scientist brother's "weather stick," a device that enabled him to control the weather. Much like any other self-respecting comic-book villain, once in possession of a weapon giving him mastery over the fundamental forces of nature, he immediately adopts a colorful costume, calls himself the "Weather Wizard," and sets upon robbing banks and vandalizing police stations. The finale of the story, as shown in fig. 9, comes when, "with a tremendous surge of speed, Flash slams toward his foe so fast that the air in front of him piles up into a wave-front and a split-instant later strikes Mardon like a solid sheet

Fig. 9. Panels from "The Challenge of the Weather Wizard" (Flash # 110) demonstrating that the faster one moves, the harder it is to get the air out of the way.

of glass." This is indeed a physically accurate consequence of the Flash's supersonic velocity, and is the source of the "sound barrier" that bedeviled fighter pilots in the 1940s and 1950s.

The density of water is much greater than that of air—water molecules are in contact with one another, while there are large open spaces between air molecules. It is therefore even more difficult to move through water when traveling at high speeds. But for the Flash, when running on top of the water's surface, this is a good thing. Just as someone is able to water-ski if he or she is towed at a high speed, the Flash is able to run faster than the response time of the water molecules. As his foot strikes the water's surface at speeds greater than 100 mph, the water is not able to move out from underneath his boot fast enough and instead forms a shock front, similar to the shock front that forms in front of a supersonic airplane. At these high speeds, the water acts more like a solid than a liquid beneath the Flash's fleet feet (to test this, try rapidly slapping a pool of water), and therefore his oft-shown ability to run across bodies of water is indeed consistent with the laws of physics. In fact, at the speeds at which he typically runs, it is practically impossible for him to *not* run across the water's surface. However, in order to acquire forward momentum, the Flash must push back against the water. That is, even if the water does behave like a solid under the rapid compression under his feet, would the Flash be able to obtain traction in order to run? One way he could accomplish this is by generating backward propagating vortices under his feet, thereby gaining a forward thrust under Newton's third law. This mechanism was recently proposed as the means by which young water-strider insects propel themselves along the water's surface. Here again comics were ahead of the curve. The Flash's ability to run across a body of water was likened to a rapidly skipping shell skimming over the water in *Flash* # 117, more than thirty years before scientists understood the water strider's method of locomotion.*

Once the Flash has moved the air before him out of his path, he leaves a region of lower density air in his wake. Compared to the surrounding air at normal density, this lower density trail behind the Flash can be considered a partial vacuum. Air rushes in to fill any vacuum, and anything standing in the way of this rushing air behind the Flash will be pushed into the wake region. The faster he runs, the greater the pressure difference between the air behind him and the surrounding air, and the larger the force as this pressure imbalance is righted. This effect is noticeable even for slower-moving objects, such as when a subway train enters a tunnel. The enclosed geometry of the tunnel accentuates the updraft behind the departing train, pulling loose newspapers and litter in its wake. Lacking a confined space, the Flash can generate a low-pressure region that can slow the descent of falling people, cars, or giant bombs, or as shown in fig. 10, help detain and levitate a crook using a vortex created by running in a circle.

Returning to the topic of the speed of sound in air, whenever the Flash runs faster than a velocity of 1,100 feet/sec (or 750 mph), his communication with others must become visual only. Anyone standing behind him or even at his sides would not be able to talk to him, as the Flash would outrace their sound waves trying to reach him. The Flash generates pressure waves that create a "sonic boom" at these speeds (more on this in the next chapter). Just such a crash heralded the first appearance of the Flash in

---

* There are important exceptions to this general principle that viscosity increases with speed, such as with tomato ketchup. When rapidly stressed, the viscosity of ketchup decreases, while as just argued a sharp shock to water increases its resistance to flow. This is why fast, hard raps to the bottom of a ketchup bottle momentarily reduce its viscosity and speed up its egress from the bottle.

Fig. 10. When the Flash runs at high velocity in a circle, he leaves a low-pressure region in his wake, which makes it easy to bring Toughy Boraz (yes, that's actually his name) and his stolen loot to police headquarters. From Flash # 117.

Showcase # 4. Of course, for anyone standing in front of him, the Flash outracing the sound waves would not be a problem, but there would still be a barrier to communication. Sound waves need a medium in which to propagate. What we term "sound waves" are actually variations in the density of the medium, alternating regions of expansion and compression. In a dilute medium, such as air, there are large spaces between molecules, which makes it harder for density variations to propagate, compared with water, steel, or the thin walls of an apartment. Roughly speaking, the denser the medium, the faster the sound travels, which is why in old western movies a character would put his ears to the railroad tracks to determine whether a train, too distant to see, was approaching. He could hear the train's vibrations through the steel rail much sooner than if he waited for the same noise to reach him through the air. Taken to its extreme, sound does not travel at all in the most dilute medium of outer space, which has a density of one atom per cubic centimeter, as compared to air with a density of twenty million trillion atoms per cubic centimeter at sea level on Earth.

Even when he can hear someone talking to him, their speech will have a high and tinny quality to the Flash. The distance between adjacent compressed (or expanded) regions in the medium in a sound wave is labeled the "wavelength" of the sound wave, which is related to the pitch that we hear. The pitch or frequency measures the number of complete wave cycles that pass a given point per second. Long wavelengths have low pitches (think of the deep tones from a bass violin, where the length of the strings is related to the wavelength of the sounds they can produce) while shorter wavelengths are heard as higher pitches. As the Flash runs, even if he does not outrace the sound wave, his high-speed motion affects the pitch that he hears. Let's say he runs toward someone who is yelling a warning to him. The sound waves have some wavelength, which marks the average distance between adjacent compressed or expanded regions. If the Flash is standing still when these alternating density regions reach him, the tone he would hear would be determined by the wavelength originating from the speaker. But as the Flash runs, one region of compressed air reaches him, and as he is running toward the speaker, the next region of compressed air reaches his eardrum sooner than it would if he were standing still. The Flash thus hears a smaller wavelength and hence a higher frequency due to the fact that he is running toward the source of the sound. The faster he runs, the greater this shift in the wavelength and frequency of the detected sound.

This phenomenon is known as the Doppler effect, and if one knows the wavelength of a stationary source of waves, and measures the wavelength of the detected waves, with a moving detector, one can determine the speed of the detector. Alternatively, if one sends out a wave of a known wavelength, and it bounces off a stationary target, it should return with the exact same wavelength. If the target is moving toward the source, the reflected wave will have a shorter wavelength, while if the target is moving away from the source the detected wavelength will be longer. Doppler radar, as often discussed on the Weather Channel, involves detection of this wavelength shift, which enables meteorologists to calculate the wind velocity in an approaching storm front.

This is also the basic premise underlying radar guns, which use radio waves of a known wavelength. From the shift in wavelength of the reflected wave to the transmitted wave, they can determine the velocity of the object (such as a thrown baseball or a speeding automobile) that reflected the waves. Of course, for this to work, the waves striking the object must be smoothly reflected, like light from a mirror, and head in a straight line back to the source (and detector). Were you to wrap your car in crumpled tin foil, the radio waves would scatter in many different directions, making an accurate determination of velocity with a radar gun very difficult (which explains the difficulty the authorities have in bringing Spud Man and his Baked Potato-Mobile to justice).

The faster the target is moving, the greater the wavelength shift, and the higher the pitch of the detected wave. If the Flash were to run at 500 mph toward someone who was using a normal speaking voice with a pitch of about 100 cycles per second, the sound waves reaching the Viceroy of Velocity's ears would be shifted to 166 cycles per second. In order for the pitch of the sound waves reaching the Flash to be greater than 20,000 cycles/second, the upper range of human hearing, the Flash would have to run toward the speaker with a speed greater than 150,000 mph (that is, 0.02 percent of the speed of light).

The Flash's preferred technique for stopping bullets is also consistent with Newton's laws of motion. You don't need to be bulletproof when you can outrun a bullet. But what about the innocent bystanders caught in the line of fire? Fig. 11 shows a physically accurate use of super-speed in this situation. As narrated by one such potential victim in *Flash # 124*, "the amazing speedster merely made his hand travel at the same speed as the bullets whizzing at him, and with a sweeping motion plucked them right out of the air before they could harm him." That is, the Flash would first match his velocity to the bullets so that the relative

Fig. 11. *The Flash demonstrates that Impulse and Momentum principles are still important, even when you can run as fast as a speeding bullet. From* Flash # 124.

speed between him and the bullet is zero. Just as one can easily pick up a book or a cup on an airplane in flight if it is not moving relative to you, the Flash is then able to pluck the bullet out of the air, since he is also moving at approximately 1,500 feet/sec or more than 1,000 mph in the same direction as the bullet. An "editor's note" in *Flash* # 124 correctly points out that "Flash's action in stopping the bullets is similar to that of a baseball fielder who stops a hot grounder by letting his glove travel momentarily in the same direction as the ball."

As discussed chapter 3, the problem with high velocities is not the speed but the deceleration. For Gwen Stacy, the braking time was very short, so the stopping force was large. The boxer rolling with a punch, as noted earlier, deliberately increases the contact time, in order to minimize the stopping force. The Flash, as the editor's note correctly points out, is applying the same principle in this situation. In addition to being able to run at amazing speeds, Barry Allen also apparently gained the ability to withstand crushing accelerations every time he sped up or slowed down. Thus, when the Flash stops running, the bullet he is holding stops as well, and he can then drop the slug at the gunman's feet for dramatic effect.

LIKE A FLASH OF LIGHTNING—

SPECIAL RELATIVITY

IN THE PREVIOUS CHAPTER I mentioned the sonic boom that the Flash creates whenever he runs faster than the speed of sound. Why is there a "boom" when an object moves at or faster than the speed of sound? And how does this help us understand Einstein's Special Theory of Relativity?

First let's consider the boom, and then get to Einstein. Imagine yourself standing out in the countryside, and the Flash is running toward you *at* the speed of sound—that is, at one fifth of a mile per second. If he starts ten miles away from you, he'll reach you in fifty seconds. When he is ten miles away, he says, "Flash" and when he is only five miles in front of you, he says, "Rules." What do you hear? If the Flash was running slower than the speed of sound, then the "Flash" he spoke would reach the five-mile mark before he did, and then he would utter "Rules" while the "Flash" was about to reach your ears. You would clearly hear "Flash Rules," followed a short while later by the sound of the Scarlet Speedster running past you.

If the Flash were instead running *faster* than the speed of sound, then he would arrive at the five-mile point *before* the sound he emitted at the ten-mile mark. He would then say "Rules" and continue on toward you. Since the "Rules" has less distance to travel, it would reach you before the "Flash," so you would hear the words in the reverse order that they were spoken—that is, to you it would sound like he'd said, "Rules Flash." This backward speech would not reach you until after he had passed you. Running faster than the speed of sound, he can cover the distance from five miles away to you in less time than the sound waves.

If he were running at exactly the speed of sound, then when he yelled "Flash" at the ten-mile point it would reach the five-mile mark at the exact same instant as the Viceroy of Velocity himself did. When he then says "Rules," it takes off from the five-mile point toward you at the same moment that the "Flash" does, so that they both reach your eardrums at the same moment, twenty-five seconds later. You don't hear "Flash Rules" or "Rules Flash" but the two words superposed at the same instant. Sound is a pressure wave, so the waves from the two words add up and create a larger vibration than if heard separately. The Flash would not even have to be talking or making noise as he advanced toward you, but simply the disturbance as he pushed the air out of his way would create a pressure wave that you would hear as a thunderous roar (or a "sonic boom") at the exact instant the Sultan of Speed cruised by. If the Flash ran faster than the speed of sound, this disturbance would still be created. In this case he would race past you in relative silence, and then later on the sonic boom, traveling at the speed of sound would eventually reach you, with explosive consequences. (The "Rules Flash" he spoke in the previous paragraph would be lost in the sonic crash.) The "crack" of gunshot, or of Catwoman's whip, are mini sonic booms created by the bullet or the tip of the whip moving faster than the speed of sound in air.

The danger posed by the indiscriminate creation of sonic shock waves by the Flash is recognized by current comic-book writers. In *DC: The New Frontier*, a 2004 revisiting of DC Comics' Silver Age heroes, set in the late 1950s when they historically made their first appearances, the writer Darwyn Cooke describes a scene where the Flash races from Central City (hazily located in the American Midwest) to Las Vegas, Nevada. In caption boxes that describe the Flash's thoughts as he runs cross-country, he tells us, "I wait until I'm clear of the city limits before I hit the sound barrier. That's one I figured out the hard way a few times. Flying glass and pedestrians don't mix." Indeed they don't, as graphically demonstrated in *Flash* # 202 (vol. 2). In this story our hero has lost his memory and, in civilian clothes, does not realize that he possesses super-speed. Acting instinctually when mugged by a street gang, his high-velocity movements blow out every window on the block and cause massive structural damage to the surrounding buildings.

Regardless of the order in which you hear what the Flash says, whether it's "Flash Rules" or "Rules Flash," if you have keen eyesight and can read lips you can be sure of the order in which the Flash actually said these words. This agreement is due to the fact that the light reflecting from the Flash travels much faster than sound (186,000 miles in one second, compared to one fifth of a mile per second). This is how we are able to determine the distance a thunderstorm is from us by comparing the timing between the lightning and the sound of thunder.*

But what if the Flash were running close to the speed of light? All sorts of strange things happen concerning length, time, and mass for objects moving near light speed, as explicated by Albert Einstein in 1905 in his Special Theory of Relativity (it's called "special" because it deals with objects moving at a constant speed, while the "General" Theory of Relativity, developed in 1915, concerns accelerating objects). This is neither the time nor place to go into a full discussion of Relativity. A fair treatment of the topic would overwhelm the present book. But I do want to mention a basic point about traveling near the speed of light that won't take too long, that we'll build on when considering the connection between electricity and magnetism in Chapter 17.

The Special Theory of Relativity can be boiled down into two statements that appear simple, but contain a wealth of physical insight. They are: (1) nothing can travel faster than the speed of light (sorry, Superman and Flash), which is the same speed for everyone, no matter how fast he is moving, and (2) the laws of physics are the same for everyone, regardless of whether you are moving or not. The first point is the really weird one. If the Flash is running as fast as a speeding bullet, to us the bullet is traveling at 1,000 miles per hour, while to the Flash, running in the same direction at the same speed, the bullet appears stationary (which is why he is able to, "with a sweeping motion," pluck it from the air

---

* The flash of lightning originating from the thunderhead covers a distance of one mile in roughly five millionths of a second (we are physiologically unable to detect events happening that quickly; for us it is instantaneous), while the sound of thunder that is created simultaneously takes nearly five seconds to reach us. Counting the number of seconds between the two events, and making use of the fact that sound takes five seconds to cover one mile, allows us to easily calculate the distance of the thunderstorm from us.

so easily). But the speed of light is 186,000 miles per second to both you standing still and to the Flash, regardless of how fast he runs. Even if he is racing at half the speed of light—at 93,000 miles per second—the speed of light relative to him is *not* 93,000 miles per second, but still 186,000 miles per second, the same as for you standing on the street corner. How can this be?

When the Flash runs toward you, from the Flash's point of view it is as if he is stationary but you are racing toward him. The Special Theory of Relativity states that for both you and the Flash, you must agree that the speed of light is 186,000 miles per second. In order for this to be true, Einstein argued that from your point of view, the Flash will appear thinner (that is, his length in the direction he is running will appear compressed) and time will seem to pass slower for him than for you. From the Flash's point of view, a yardstick he holds would still be one yard long, and his watch keeps time just as it always had, but to him it is you who are moving rather than him, and he will make similar determinations about you (your length will be shortened and time will move slower for you, as it appears to the Flash). This is because in order to measure the length of a yard-stick the running Flash is holding, for example, you have to consider the front and back ends of the stick, and clock the times when they pass a given point. For two people moving relative to each other (say, a running Flash and a stationary observer) it becomes impossible for them to agree as to whether or not two things happen at the same time if they are separated in space and time.

Information cannot travel faster than light, so there will always be a discrepancy as to the order in which events occurred. In order to balance out so that the one thing everyone agrees on is the value of the speed of light, it will appear that lengths are shortened in motion and that time passes more slowly. Comic books frequently blow this when dealing with characters who can travel at the speed of light (such as Negative Man of the Doom Patrol and Captain Marvel—not the "Shazam" guy, or the late 1960s–early 1970s officer of the Kree military, but an African American heroine in the Avengers in the late 1980s who could transform her body into coherent photons of light). The other characters in the story should not be able to see these heroes, as there's no way for light to be scattered from them if they are moving as fast as light. At best they might appear as a spark of lightning from a great distance away, but would be invisible when nearby.

The highest velocity the universe allows is the speed of light. As the Flash runs faster and faster, one would think that he should be able to pass this limit, but this can't happen. To explain this behavior, from the point of view of a stationary observer, it must be that the faster he goes, the harder it becomes for him to further accelerate. From Newton's second law (Force is equal to mass times acceleration), we know that if the force that his running shoes apply stays constant but there is no corresponding acceleration, it must be because his mass has increased. So in addition to time seeming to be slower and lengths appearing to shrink, the mass of the running Flash will seem (to we stationary slowpokes) to grow, the faster he runs. This is occasionally referred to in the comic books. In *JLA* # 89, the Flash has to move the entire population of 512,000 men, women, and children in Chongjin, North Korea, away from the imminent blast of an atomic bomb in a fraction of a second. In order to accomplish this feat, he must move at speeds very close to the speed of light. The relativistic consequences of his high speeds are alluded to once he has saved the town and collapses to his knees on a hilltop. As described in a caption box, "as his body sloughs off the screaming aftereffects of near light travel, eyes of almost infinite mass turn towards the blaze engulfing Chongjin." The relativistic gain in mass will in fact only occur while the Flash is running. By the way, it was the realization that an increase in the kinetic energy of an object is directly connected to a rise in its mass that led to Einstein's most famous equation of $E = mc^2$.

In addition to being able to run at amazing speeds, the Flash is said to possess total control over every one of his molecules' motions. This vibratory control was put to use in *Flash* # 116, and many times since. Matching his body's vibrations to the vibrational frequency of the atoms in a wall, it was argued by the writers of *Flash* comics that he could pass through a solid wall, without harm to either himself or the wall. However, it is not true that the only reason we cannot walk through solid walls is that we vibrate at a different frequency from the atoms in the wall. In fact, as will be discussed later on, the average rate at which our atoms vibrate is simply a reflection of our temperature. Our bodies are typically roughly within 20 percent of the temperature of a wall, so our atoms' vibrational frequency is already pretty well matched to those in the wall.

Nevertheless, there *is* a quantum mechanical phenomenon termed "tunncling" that allows one object to pass through a solid barrier without disturbing either the barrier or itself. A discussion of this quantum effect is premature, as we are still concerned with Newton's three laws of motion and classical physics. We will therefore defer a consideration of this Flash feat to chapter 22.

7

IF THIS BE MY DENSITY–

PROPERTIES OF MATTER

BEFORE DR. HENRY PYM began moonlighting, fighting crooks, and capturing communist spies as Ant-Man, he was a fairly typical biochemist. In his first appearance in "The Man in the Ant Hill," Pym was shown struggling with the bane of a modern scientist's life: funding! As we learn in a flashback, at a recent scientific convention a panel of scientists went beyond merely rejecting Pym's request for financial support of his search for a shrinking potion, and took the additionally cruel step of personally taunting him. "Bah! You're wasting your time with your ridiculous theories," jeers one professor, "but they never work!" Another counsels: "You should stick to practical projects!" To which Pym replies, "No! I'll work only on things that appeal to my imagination . . . like my latest invention!" I should point out that two aspects of this exchange ring particularly true: namely (1) there continues to this day a tension at universities and research laboratories, between research that is motivated by the search for practical applications and those investigations that are "curiosity-driven," and (2) unlike the public at large, scientists routinely use the expression "Bah!" in everyday conversation.

Pym's first accidental exposure (and you would not be far wrong to conclude that nearly all superheroes are fairly accident-prone, at least when it comes to gaining their powers) to his shrinking potion led to a harrowing adventure inside an ant hill, reminiscent of the 1954 science-fiction story "The Incredible Shrinking Man." At

Fig. 12. *The opening page of "The Return of the Ant-Man" from* Tales to Astonish # 35, *in which we meet the costumed superhero alter-ego of Dr. Henry Pym for the first time.*

the end of this story Pym regains his original height through the application of the growth potion, and, once normal size, pours them both down the sink. Realizing that the potions are "far too dangerous to ever be used by any human again!" he vows, "From now on I'll stick to practical projects!" What Pym considers more practical than developing a reversible miniaturization process is left to the reader's imagination.

Dr. Pym's vow remained unbroken until the sales figures came in for *Tales to Astonish* # 27. As shown in fig. 12, by *Tales to Astonish* # 35, the good doctor was back in "The Return of the Ant-Man" (though in the prior story in *Tales to Astonish* # 27 he had

never referred to himself by that title), having replicated his shrinking potion and designed a snazzy red-and-black jumpsuit and a
"cybernetic" helmet that enabled him to electronically communicate with ants. Given that ants actually communicate with one
another through the sharing of pheromone chemicals that they
excrete, we won't look too closely into how Pym's helmet might
actually function. So arrayed, scientist Henry Pym fought common criminals, communist spies (it was the early 1960s, after all),
invading aliens, and bizarre supervillains such as the Porcupine
and Egghead as the astonishing Ant-Man. It seemed as if no evil-
doer could win in a face-off (though not a literal one given the size
differential) with a crimefighter whose superpower was that he
was only a quarter-inch tall.

Adventures involving characters reduced to the size of insects
have been a staple of science-fiction movies and comic books for
at least fifty years. Yet here we are in the twenty-first century, and
we've still not achieved this radical form of weight reduction.
What's the holdup?

After all, it truly seems like every other day brings a news report confirming the equation: **science fact equals science fiction
plus time**. Robots assemble automobiles or vacuum your apartment; a computer has beaten the world champion in a chess
tournament; and therapeutic cloning promises to alleviate many
devastating diseases and medical conditions. Man has gone to
the moon, walked upon its surface, and returned safely to Earth,
not just once but several times—and travel to other planets, at
least within our own solar system, has been accomplished, if only
by unmanned craft so far. Scientific papers have even been published in prestigious physics journals discussing the construction
of "time machines," whose operation involves the concept of
"negative energy." (This "negative energy," when combined with
the mechanics of wormholes—a concept developed in the General Theory of Relativity—has been postulated as providing a
theoretical mechanism for warp speed; that is, faster-than-light
travel.)*

---

* This "negative energy" is associated with squeezed quantum states, and is
beyond the scope of this book. And no, your brother-in-law cannot be considered a vast, untapped source of negative energy.

On the technology end of futuristic projections, the handheld communicators from *Star Trek* are now an everyday item, and in fact some cell phone models, with digital image storage and transmission, and Internet access, exceed the imaginations of *Star Trek* writers from the 1960s (the two-way wrist TV communicators of the Dick Tracy comic strips are not far off). *Star Trek's* "tricorders"—handheld devices the size of a hardcover book that enable chemical and biological analysis—may soon be available for purchase: PDAs are already common, and the technology to perform "DNA analysis on a chip" and other functions is in development. From flat-panel television screens to microwave ovens and magnetic resonance imaging that provides three-dimensional views of the human interior, we are indeed living in the World of Tomorrow, even if we still lack portable jet packs and robotic butlers.

Yet despite all the fantastic marvels and concepts that are either already here or seem within our grasp, we still can't shrink or enlarge people at will. Compared to miniaturization, warp drives and time travel are right around the corner. However, back in the 1960s, it was shrink-rays that were promised by our comic books and movies, coming soon to a top-secret underground military scientific laboratory near you.

The 1966 science-fiction film *Fantastic Voyage* described the adventures of a surgical team and a mini-sub that is miniaturized to the size of a bacterium and injected into a scientist's bloodstream, to remove a blood clot in his brain that is inoperable from the outside. Before the movie begins, a title card appears, reading: THIS FILM WILL TAKE YOU WHERE NO ONE HAS EVER BEEN BEFORE, NO EYE WITNESS HAS ACTUALLY SEEN WHAT YOU ARE ABOUT TO SEE. BUT IN THIS WORLD OF OURS WHERE GOING TO THE MOON WILL SOON BE UPON US AND WHERE THE MOST INCREDIBLE THINGS ARE HAPPENING ALL AROUND US, SOMEDAY, PERHAPS TOMORROW, THE FANTASTIC EVENTS YOU ARE ABOUT TO SEE CAN AND WILL TAKE PLACE. Three years later man did indeed walk on the moon, and it is certainly true that compared to thirty years ago, the most incredible things are happening all around us. Yet we have much longer to wait before a team of doctors can make this ultimate house call. What is the insurmountable barrier that prevents an aspiring Dr. Henry Pym from radically changing size?

The reason that miniaturization is physically impossible (as far as we know) is that matter is made of atoms, and the size of an atom is a fundamental length scale of nature, not open to continuous adjustment. As discussed in Isaac Asimov's novelization of *Fantastic Voyage*, to make something smaller requires either: (1) making the atoms themselves smaller, (2) removing some (large) fraction of its atoms, or (3) pushing the atoms closer together.

* * *

First let's consider the size of the atoms. In cartoon representations of atoms, in DANGER! RADIOACTIVITY! warning signs for example, the orbits of electrons around the nucleus are represented as elliptical trajectories, like those the planets make about the sun in our solar system. We would mark the "size" of our solar system as the distance from the sun at its center to the outer boundaries of the orbits of the planets, and similarly the "diameter" of an atom would be determined by the range at which the electrons buzzed around the nucleus. The typical size of an atom is about a third of a nanometer, where one nanometer is one billionth of a meter (a meter is roughly 39 inches long). This seems small, and it is: Looking along the cross section of a human hair, about 300,000 atoms lie end to end across its width.

Every atom has a nucleus consisting of a certain number of positively charged protons and a comparable number of uncharged neutrons. In addition to the positively charged protons, the atom contains an equal number of negatively charged electrons. If oppositely charged objects attract one another, then why don't the positively charged protons pull the negatively charged electrons toward them, until the electrons sit on the nucleus? Well, they would, if the electrons were standing still. After all, as discussed in chapter 2, the Earth and moon pull toward each other through their mutual gravitational attraction, and the moon's orbit is such that its distance from the Earth and its speed exactly balance the inward gravitational pull. Similarly, the electrons reside in "orbits" around the nucleus at the center of the atom. Interestingly enough, all atoms are roughly the same size, to within a factor of three. The number of protons in the nucleus is countered by an equal number of "orbiting" electrons. Heavier atoms have more protons that pull the electrons more strongly toward the nucleus, but more electrons mean there is more repulsion between the negatively charged electrons, trying to get away from each other. This balancing act results in a "size" of the atom that is roughly twenty or thirty billionths of a centimeter.

I must note, for reasons that we will get into in section three, that this picture of electrons sweeping out precise elliptical orbits in an atom is not correct. Rather, quantum mechanics tells us not where the electrons are, but provides a mechanism for calculating the probability of finding an electron at a certain distance from the nucleus. It is the distance for which the probability of finding an electron is largest (the densest region of the "probability cloud") that is referred to as the "radius" of the atom, and is associated with the atom's size. The expression for the most probable radius of the atom depends only on such terms as the mass of the electron, its electric charge, the number of positive charges in the nucleus, and a fundamental constant of the universe **h**, known as Planck's constant (the value of which determines the magnitude of all quantum phenomena). We will discuss **h** in more detail in section three, but for now all we need worry about is that **h** is a fixed number, just as the mass of an electron or the magnitude of its electric charge is in the expression for the atomic radius. Once the number of positive charges in the nucleus (the quantity that determines what element we are dealing with) is set, there's nothing to change. The size of an atom is determined by a collection of fundamental constants and is *not* open to adjustment.

In Isaac Asimov's follow-up to his novelization of *Fantastic Voyage*, titled *Fantastic Voyage II: Destination Brain*, the mechanism proposed to enable miniaturization involved the creation of a "local distortion field" that somehow manages to change the value of Planck's constant. If **h** becomes a tunable parameter, a factor of ten reduction in its value would shrink the size of an atom to one hundredth of its present size. Needless to say, we have no idea how to begin to accomplish this in the real world, which is, after all, why **h** is considered an invariable constant. Our lives would be profoundly different if we ever discovered a way to change the fundamental constants of nature so that the speed of light or the charge of an electron become open for our adjustment.

Until that day comes, these constants are just that, and because the radius of an atom is described by said constants, the size of an atom cannot be changed. We therefore cannot make atoms themselves smaller—at least, not without also changing the type of universe we live in.

What about the second suggestion for size reduction—that is, removing a fraction of the atoms in an object? Everything is made of atoms; consequently removing some of them should make an object smaller. Certainly the reduction in the size of electronic devices (the "crime lab on a chip" alluded to earlier) suggests that some objects can be made with less material and still retain their functionality. Problems arise however with complex, living things, for which the removal of a significant number of atoms would have serious consequences. To go from a height of six feet to six inches is a reduction in height by a factor of twelve. Of course, people are three-dimensional, and a factor-of-twelve reduction in the width and breadth would be required as well. To accomplish this by removing atoms (assuming that one could do this, had a safe place to store them, and could replace them later when you wanted to regain your original height) means that you would get to keep only one atom for every 1,728 atoms subtracted. Even assuming that this removal is performed uniformly—such that the same fraction of atoms is removed from all of your cells—biological functionality would be lost or at least severely compromised.

Consider the neurons in your brain. It is a myth that man uses only 10 percent of present brain capacity; evolution theory argues against such a monumental waste of available resources. If neurons could be smaller and still fulfill their role in the brain, then there would be a strong competitive advantage to any mutation in this direction. Not only would it require fewer atoms to build a person (so the demands for raw materials through food would be greatly reduced) but one could have many more neurons and hence synaptic connections if our brains retained their current size. The typical neuron has a width of approximately one thousandth of a centimeter, and this is true whether one is considering an ant's neuron or a human's. People are smarter than ants (on average—I'm sure we can all think of a counterexample from our personal experiences) because we possess roughly four hundred thousand times more neurons and a correspondingly larger number of synaptic connections, not because our neurons are a thousand times bigger. Remove 99 percent of the atoms, and you can make your body's cells 99 percent smaller, but they won't work as intended.

Finally, what about the third possibility—shrinking a person by making him denser by pushing the atoms closer together? Unfortunately, this is not a successful strategy for miniaturization either, for the same reasons that Krypton could not simply be a planet 15 times denser than Earth. Atoms in most solids are already tightly packed. In addition, thanks to the repulsion of negatively charged electron probability clouds around each atom, they are fairly rigid objects and resist being squeezed together, as would a child's collection of marbles in a box. When a volume is filled with marbles, nearly all of the container's available space is taken up by the marbles. With few exceptions, every marble is in physical contact with several of its neighbors. Certainly there are open gaps between the marbles, but these spaces are not big enough for us to add more than possibly a few percent more marbles. If the marbles are hard spheres and not compressible, then squeezing on the walls of the container will not lead to a significant reduction in its volume. Reducing the size of the container by a factor of ten would require pressures that would crush the marbles. Trying to shrink a person by applying similar pressures would result in comic-book stories that would be both brief and messy and almost certainly would not garner approval by the Comics Code Authority.

* * *

Well, if miniaturization is so difficult, how does Henry Pym, also known as the Ant-Man, accomplish it? As told in *Tales to Astonish* # 27, biochemist Dr. Henry Pym had devoted years to discovering a potion that would shrink any object, until treated with an antidote growth serum. Later, Pym would convert his shrinking potion into an easy-to-swallow pill form. When it eventually became necessary to explain how he was able to shrink other objects, such as his costume, helmet, and weapons that he carried, it was revealed that he had developed a small generator of "Pym particles" that were able to increase or decrease an object's size. No explanation has ever been put forth for how these potions or Pym particles actually work, and their physical basis must fall under the "miracle exemption" we frequently invoke when considering the source of a hero's superpowers.

8

CAN ANT-MAN PUNCH HIS WAY OUT OF A PAPER BAG?–

TORQUE AND ROTATION

EVERY COMIC-BOOK HERO has some Achilles' heel, and Ant-Man's was only a millimeter big. There are certain obvious drawbacks for a hero who is the size of an ant. For example, just as Superman is susceptible to Kryptonite, Ant-Man has to be ever vigilant against the more common hazard of being stepped on. In addition, his stride being only a few millimeters, he requires thousands of steps to cover the same distance he could walk in one step while normal height. The time required for him to walk a few feet would therefore increase correspondingly. No doubt this was his motivation for frequently hitching rides atop carpenter ants. The fact that he could ride on top of an ant without crushing it indicated that Ant-Man's mass decreased along with his size, implying that his density remained constant as he shrank. (Remember that density is the mass of an object divided by its volume; if the volume decreases by a factor of one thousand, and the mass is reduced by an identical factor, their ratio and hence the object's density is unchanged.) Pym made good use of his reduced mass and constructed a spring-loaded catapult that could shoot him across town. Of course, as we've discussed at length in chapter 3, it's not the journey but the stopping that is problematic. In order to avoid a messy finale to his trajectory, Pym called upon his special rapport with ants and used his cybernetic helmet to instruct hundreds of them to form a living air bag to cushion his landing. Ant-Man's kinetic energy would be distributed among the many, many ants so that no one insect would suffer overmuch for its participation in breaking his fall.

If Ant-Man is such a lightweight that he could be propelled across several city blocks by a coiled spring and not harm the ants that stopped his motion, then how is he able to disable such foes as the Protector or the Hijacker, or meet "The Challenge of Comrade X"? In particular, how is Ant-Man able to punch his way out of a vacuum cleaner bag (shown in fig. 13), as thrillingly rendered in *Tales to Astonish* # 37, or swing a crook overhead using a nylon lariat in the very next issue? As explained in *Tales to Astonish* # 38, Henry Pym retained "all the strength of a normal human," even when ant-size. Not to nitpick, but the average normal-size

Fig. 13. *A scene from "Trapped by the Protector" from* Tales to Astonish # 37, *in which it is demonstrated that Ant-Man is both as light as an ant (and hence easily captured by a vacuum cleaner) and as strong as a normal-size human (and consequently able to punch his way out of the vacuum bag).*

human, not to mention the average biochemist, would be hard-pressed to be able to swing a full-grown man overhead, even using a "practically unbreakable" nylon lasso. But leaving that issue aside, what does it mean to say that Ant-Man has the strength of a normal-size person, such that he can break a vacuum bag, but only the mass of an ant, such that he is easily sucked up by the vacuum cleaner in the first place? Perhaps the more basic question is: Why do you have the strength that you do, such that you can easily lift a 20-pound object, but struggle with one weighing 200 pounds and can not possibly lift 2,000 pounds? Our strength comes from our muscles and skeletal structure that make up a series of interconnected levers. It turns out that these levers are not that well suited to lifting things.

Let us stipulate that, of its many definitions, by "strength" we mean the ability to lift an object. Mankind's ingenuity has led to the development of a wide range of machines to perform tasks such as heavy lifting. One of our earliest inventions created to lift objects is the simple mechanical device of a lever. Many of us first encountered a lever as children in the form of a playground's "teeter-totter" or seesaw, consisting of a horizontal board supported by a fulcrum point placed beneath the midpoint of the board. When seated at one end of the seesaw, you are able to lift a playmate high in the air, through the mechanical advantage of the lever. With the fulcrum placed at the exact middle of the board, you can only lift a mass roughly equal to your own. If, however, the fulcrum point is placed much closer to one end, then a small child can lift a full-grown adult, provided the adult sits at the end of the seesaw nearer to the fulcrum point. This is because seesaws, and levers in general, don't balance forces but rather "torques."

If a *force* is able to push or pull an object in a straight line, then a *torque* is a measure of the ability to rotate an object. A torque is mathematically defined as the product of the applied force and the distance between the force and the point where the object is to be rotated. While both "torque" and "work" are defined mathematically as the product of force and distance, in the case of work the distance is the displacement of the object—that is, the distance over which the force pushes or pulls the object (more on work in chapter 11). The force must be acting in the same direction as this distance in order to change the object's energy. In contrast, for a torque the force is at a right angle to the separation between the applied force and the point about which it is to be rotated. This distance is sometimes referred to as the "moment arm" of the torque. For a given applied force, the larger the force's distance from the point where the object is to be rotated, the greater the torque.

This is why doorknobs are placed at the end of the door as far away from the hinges as possible. Try closing a door by pushing it at the end that's immediately adjacent to the hinges, and then apply the same force to the other end, where the doorknob is located. The same force is used, but increasing the moment arm by increasing the distance from the push and the hinges, magnifies the torque, and produces an easier door closing. A wrench is another simple machine, which amplifies a force applied at one end to produce a rotation at the other. When trying to loosen a particularly stubborn nut, one sometimes makes use of a "cheater," basically an extension arm for the wrench by which the moment arm, and hence the applied torque, can be increased when the available force that can be applied is already at a maximum. Returning to the example of the seesaw, a small child is able to lift a full-grown adult only when the fulcrum of the seesaw is placed near the adult's end (in a playground seesaw, the adult usually sits closer to the fixed fulcrum in the center). In this case the moment arm for the child is increased, and the torque she applies is large enough to rotate the adult up into the air, which the child could not accomplish without the mechanical advantage provided by the lever.

Levers also play a role in determining the strength of Ant-Man's tiny punch. Our arms are able to lift and throw by making use of the principle of levers. An object, let's say a rock, is placed on one end of the lever, which we'll call a "hand." A force is exerted by the compression of a biceps muscle, causing the other end of the lever (the forearm) to move down, which in turn raises the far end of the lever—that is, the hand holding the rock. The biceps pulls the hand upward, while when we need to lower the rock, the triceps contracts and in so doing pushes the hand back downward. Muscles can only contract and pull; they cannot push. Consequently an ingenious series of levers, consisting of muscles attached to various points of our skeletal structure, has evolved in order to enable a wide range of movement. The fulcrum of the lever that is your forearm is located at the elbow. It may seem odd to have both forces applied on the same side of the fulcrum, but this type of lever is essentially the same as a fishing rod, where the force applied to one end—very near the fulcrum located near the reel—causes a rotation and consequent lifting of a fish at the other end of the rod. Your bicep applies a pulling force approximately two inches in front of your elbow, while most people's forearms are fourteen inches long. The ratio of moment arms is thus 1 to 7, which means that the force applied by your biceps is reduced by a factor of seven at the location of your hand. That's right, *reduced*— in order to lift a rock weighing 20 pounds, your biceps has to provide a lifting force of 140 pounds.

A reasonable response to this news would be: What's the point in that? Why have a lever built into your arm that *increases* the force needed to lift an object? There wouldn't seem to be any at all, and this would be exhibit A in the case *against* evolution, *if* the primary function of our arms were to lift rocks. Because the bicep is attached much closer to the fulcrum point (the elbow) than your hand, the biceps contracts two inches and the hand rises fourteen inches, due to the same ratio of 1 to 7 in moment arms. This ratio also holds when we want to get rid of said rock we are holding. In this case a muscle contraction of less than two inches produces a displacement of the hand of roughly twelve inches. This requires only 0.1 seconds to occur, and the hand holding the rock can get rid of it with a velocity of 12 inches in 0.1 second, or 10 feet per second (that is, 7 mph). This is a low estimate, and the average person can provide a much larger release velocity using other levers connecting her upper arm to her shoulder. A very, very small subset of the general population can throw baseball-size objects at speeds of up to 100 mph. And that's the point of the inverse lever in our arms—it's not intended to lift up large rocks; it's there to enable us to throw smaller rocks at high velocities. Those of our ancestors who were better rock- or spear throwers were, on average, better hunters. Being a better hunter increased one's chance of providing dinner, and that in turn increased the odds of getting a date. In this way certain hunters were able to pass these "good throwing arm" genes down to their progeny.

Meanwhile . . . I haven't forgotten about Ant-Man trapped in a vacuum cleaner bag. For the tiny crime-fighter, all length scales are obviously reduced, but the ratio of moment arms of 1 to 7 in his arms still holds for Henry Pym, regardless of whether he is ant-size or normal height. Punching involves the same muscles and similar motions as throwing, only instead of a rock one is throwing a fist. The force provided by your muscles does not depend on their length, but on their cross-sectional area. If Ant-Man is 0.01 times his normal height, then the force his muscles can provide is reduced by a factor of $(0.01)^2 = 0.0001$. If Pym can punch with a force of 200 pounds when full size, his scaled-down punch delivers a wallop of two hundredths of a pound. At his miniature size, his fist is much smaller and has a cross-sectional area of 0.0005 square inches (assuming his hand is just a millimeter wide). The pressure that his punch provides is defined as the "force per unit area," which is 0.02 pounds divided by 0.0005 square inches—or 40 pounds per square inch. This is to be compared to his normal-height punching force of 200 pounds divided by his normal-size fist's cross-sectional area of 5 square inches, for a pressure of 40 pounds per square inch. Henry Pym's punches pack the same pressure when he is ant-size as at his normal height. It appears that Ant-Man can indeed punch his way out of paper bag. In this way he serves as a role model and inspiration to all comic-book fans.

WHY BEING BITTEN BY A RADIOACTIVE SPIDER ISN'T ALL IT'S CRACKED UP TO BE

While we're on the subject of one's strength while the size of an insect, I would like to take a moment to dispel a myth concerning Spider-Man. As we have just argued, if Henry Pym shrinks at a constant density, then while the force of his punch is not as great as when he is at normal height, the pressure his fist is able to supply to an unsuspecting vacuum cleaner bag is unchanged. A common misconception is that this scaling works in both directions, so that if someone were bitten by a radioactive spider, to take a random example, then he would gain the proportionate leaping ability of a spider. That is, if a spider or flea can jump one meter high—which is roughly 500 times higher than its body height—then if a human had a comparable leaping ability, he would be able to leap to a height roughly 500 times his body height. If he stands six feet tall, this implies a jumping range of 3,000 feet! If this were indeed the case, then Spider-Man would have the Golden Age (pre-flight, pre-yellow-sun-derived superpowers) Superman beat by—and here the expression could be taken literally—a country mile. However, this is nowhere near the case. If Peter Parker did indeed gain the leaping ability of a spider, then he would be able to jump the same distance as a spider—namely, one meter. For the sake of exciting and engaging comic-book stories, it is a good thing that Stan Lee and Steve Ditko did not understand this scaling problem. Let's see where they went wrong.

What determines how high you can leap? Two things only: your mass and the force your leg muscles can supply to the ground. These two factors determine how much acceleration you can achieve as you lift off the ground. Once you are no longer in contact with the pavement, the only force acting upon you is gravity, which serves to slow you down as you ascend. So there are two accelerations we have to concern ourselves with: the initial liftoff boost that gets you airborne and the ever-present deceleration of gravity that eventually halts your rise. Once you are moving with some large velocity v the height h you will climb is given by the familiar formula from before $v^2 = 2gh$, where once again g represents the deceleration due to gravity.

There is a surprising aspect of this equation that we have not yet remarked upon, namely, nowhere does the final height that the leaper reaches depend on the mass of the person jumping! Big or small, if you start off with a velocity v and only gravity is pulling you back to Earth, then your eventual height depends only on the deceleration due to gravity g and your initial velocity v. There is a second acceleration that enters into the leap—that provided by your leg muscles at the start of the jump. And this acceleration does depend on the mass of the leaper. Using Newton's second law of motion, that **Force equals mass times acceleration** or $F = ma$, it is clear that for a given force F, the larger the person (that is, the bigger his mass m), the smaller will be his acceleration a, and the less of an initial velocity he will achieve. A smaller starting velocity means a lower height h you will be able to jump.

It's not that spiders are such great leapers that they can jump 30 times their body length. Rather, it's that small insects have tiny muscles (providing a small force), but they only have to lift an equally tiny mass to leap one meter, which just turns out to be many times larger than their size. Humans have much bigger muscles than spiders and can achieve much greater forces, but they have to lift much greater masses, so the net effect is that the height they can jump is also about one meter. Of course, some humans such as Olympic high jumpers can leap much higher than this, while most of us slugs can only jump barely more than a third of a meter (that is, one foot). In fact, for a flea to leap 200 times its body length requires a great deal of cheating on nature's part: In addition to being particularly stream-lined to minimize air drag, the flea pushes off with its two longest legs to maximize the lever arm. These are its hind legs, so in fact fleas always jump backward when they alight.

It is a natural mistake when scaling up the abilities of the insect and animal kingdom to human dimensions to assume that it is the proportions that are important, rather than the absolute magnitudes. In the nineteenth century many distinguished entomologists made the same error. As succinctly put in a footnote in the classic *On Growth and Form* by D'Arcy Thompson: "It is an easy consequence of anthropomorphism, and hence a common characteristic of fairy-tales, to neglect the dynamical and dwell on the geometrical aspect of similarity." But such misconceptions make for much more interesting fairy tales, and comic-book stories.

IS ANT-MAN DEAF, DUMB, AND BLIND?—

SIMPLE HARMONIC MOTION

THERE ARE SEVERAL SUPERHEROES whose special power is the ability to shrink. In addition to Ant-Man and the Wasp in the Marvel universe, the Atom, Elasti-Girl of the Doom Patrol, and Shrinking Violet of the Legion of Super-Heroes in DC Comics all share the ability to miniaturize themselves. The "explanation" for their shrinking powers varies, but they all share one feature in common: They will find it very difficult to communicate. I'm not referring to Ant-Man's and the Wasp's communication problems that led to their divorce, but rather the physical limitations involving conversations with anyone in the nonshrunken world.

If you're only a few millimeters tall, no one can hear you, nor will your own hearing be too keen, so you will have to depend on nonverbal means to communicate with nonshrunken people. As Ant-Man shrinks, his voice becomes higher pitched, until he has reached the size of an ant and his normal speaking voice will be at the upper range of normal-size-human hearing. At the same time his hearing threshold also shifts to the higher end of the register, such that he will miss most of what people are saying to him. To make matters worse, everything our tiny hero sees will be an out-of-focus blur. Let's see why being an "inch-high private eye," while not making you completely cut off from the outside world, will nevertheless lead to a host of problems.

First we address the basic questions: What determines the range we are able to speak and hear? And the answer, keeping with our theme of making every physics problem as simple as possible, involves the period of a pendulum. A pendulum is just a mass connected to a thin string (we'll neglect the string's mass) with the other end connected to a frictionless pivot point on the ceiling. The mass is usually assumed to be some dense sphere, such as a billiard ball or a bowling ball, but it could be a rock or Spider-Man. This mass is lifted to some height, such that the string makes a small angle with the vertical direction and, once released, the forces acting on it are: (1) gravity, always pulling straight down in the vertical direction, and (2) the upward tension in the string. The tension's direction makes an angle relative to the vertical and is continually changing as the mass swings back and forth. Part of the tension is aligned in the vertical, balancing the weight of the mass, and the rest of the tension points in the same direction in which the mass swings. This extra portion of the tension is the force that changes the mass's speed, and is responsible for the acceleration the mass experiences in its back-and-forth oscillations.

The time that a pendulum takes to move back and forth, from some high initial point, completing a full arc, and returning to its original position, is called the "period," and its motion is called "periodic." Whether it's Spider-Man swinging back and forth on his webbing, or a billiard ball attached to fishing line, the time to complete a full oscillation depends on only two factors: the acceleration due to gravity (what we called **g** in chapter 1) and the length of the string. Surprisingly, one thing that the period does not depend on (for small angles of oscillation) is the initial height at which the mass starts its motion.

Galileo was perhaps not the first person to notice that the period of a pendulum is an intrinsic property and independent of how high or low the swinging mass's starting point is, but he is properly credited for determining what controls the oscillation rate. As surprising and counterintuitive as it seems, the time necessary for a playground swing to go back and forth neither depends on how heavy or light the person sitting on the swing is, nor on how far back he starts his motion, but only on the length of chain between the seat of the swing and the top pivot point of the swing set. We are assuming that neither the swing is being pushed by a stationary helper nor the person in the swing is pumping as he moves back and forth. It is certainly true that the higher the starting position of the swing, the faster you will be moving at the bottom point of your swing, since the component of the tension in the string that deviates from the vertical (responsible for accelerating along the swing) is larger the greater the starting angle. Shouldn't this mean that it should take less time to complete an oscillation? No, because while you may be moving faster owing to your higher initial position, you have farther to travel until you reach that bottom point of the arc. The combination of the greater speed but longer distance to be traversed balances out, such that the time needed to complete the arc remains the same, regardless of the starting point. This is why a pendulum or any other device undergoing simple harmonic motion makes a great timekeeper. Two identical clocks, using either a swinging bob like that in a grandfather clock, or a coiling and uncoiling spring like that in old-fashioned pocket watches or metronomes, will keep identical time, independent of the starting push that begins the oscillation. A metronome is an upside-down pendulum, and its frequency is independent of how it is set in motion, but it is altered by changing the location of the mass on the swinging arm.

If the period of a pendulum does not depend on the starting point of the swing, why does it depend on gravity and the length of the string? It's not hard to see that the weaker the acceleration due to gravity, the less force there will be pulling on the mass and the slower will be its motion. On the moon a pendulum will take longer to complete a cycle than it does on Earth, and in outer space where the acceleration due to gravity g is zero, then the bob will never move, and the time to complete an oscillation (the period) would be infinite. Why does the length of the string enter into the pendulum's period? Because of geometry. The area swept out by the swinging mass resembles a slice of pizza pie, where the pivot point is the center of the pie and the trajectory of the bob is the pizza crust. For a full, uncut pizza the distance around the crust, referred to as the circumference, is $2\pi \times R$, where R stands for the radius of the circle and the Greek letter π is the constant 3.14159. . . . The bigger the radius R, the larger the circumference ($2\pi R$) and the bigger the length of crust on a single slice of pie. For the pendulum the role of the radius is played by the length of string connecting the pivot point to the mass. The distance the bob must travel increases with the length of the pendulum, thereby increasing the time needed to complete an oscillation.

The *frequency* of the pendulum—the number of back-and-forth oscillations it completes in one second—is just the inverse of the period, which is defined as the time needed to finish one cycle. If an oscillator has a period of 0.5 sec, so that it completes one full cycle in only one half of a second, then in one second it finishes two cycles. If the period is one tenth (0.1) of a second, then it will have a frequency of 10 cycles per second. The shorter the period, the higher the frequency. The square of the period, in turn, is proportional to the ratio of the length of string in the pendulum **l** and the acceleration due to gravity **g**. That is, the **(period)^2 = (2π)^2 × (l/g)**. To explain why it is the square of the period, and not just the period, that depends on the ratio of **l** over **g**, and why the factor of **(2π)^2** enters, would require us abandoning our "algebra-only" pledge. For our purposes the important point is that one has to increase the length of the pendulum's string by a factor of four to double the period. Conversely, shrinking the length of the string (say, by using Pym particles) decreases the period, and the smaller the period, the higher the frequency.

A human vocal cord is not a mass swinging back and forth on a string, but the beauty of the pendulum as a description of simple harmonic motion is that it captures the important physics of *any* oscillating system.* When Henry Pym shrinks down to ant-size, he reduces his dimensions by roughly 300 times. The fundamental frequency of the oscillator is, correspondingly, seventeen times bigger (that is, by the square root of 300). Normal human speech occurs at a pitch of roughly 200 cycles per second, but for an ant-size person the frequency is shifted up by this factor of 17, to 3,400 cycles per second. Our hearing range extends from 20 cycles per second at the low end up to 20,000 cycles per second, so we should still be able to hear Ant-Man, but he will have a high-pitched voice, as his chest cavity similarly shrinks. Hearing a quarter-of-an-inch-tall superhero order a supervillain to surrender in such a squeaky

---

* The periodic harmonic motion of a swinging pendulum is one of the cornerstones of much of the theoretical modeling one does in physics. So many times, in attempting to describe some complicated natural phenomenon, we begin by invoking a simple pendulum that one is tempted to paraphrase Yogi Berra and state that 90 percent of physics is "simple harmonic motion," and the other half is the "random walk."

voice, it is surprising that Ant-Man's foes do not succumb to fits of laughter rather than his tiny right hook.

Not only will his voice change as he shrinks, but Ant-Man's hearing will also be affected by his reduced height. The resonant frequency of a drum also rises as its diameter shrinks. A large bass drum has a deep, low tone, while a smaller snare drum elicits a higher pitch when struck. When Dr. Pym's eardrums shrink upon exposure to Pym particles, the frequencies that he is able to detect shift accordingly. (The physics underlying the range of human hearing is actually pretty complicated, but for our purposes we'll assume that it is determined by the eardrum.) The lowest frequency he can hear when at his normal six-foot size is roughly 20 cycles per second which, upon shrinking, becomes 17 times greater at nearly 340 cycles per second. A normal person's speech, at a pitch of 200 cycles per second, will therefore be below the range of detection for our tiny titan. For this reason, Ant-Man and his miniaturized colleagues will need to be astute students of body language as they interact with the normal world.

In addition to the change in the threshold frequency of his eardrums, as Henry Pym shrinks to the size of an ant, his hearing sensitivity will also be affected. As the vocal cords vibrate back and forth, they cause alternating compressions and expansions on the air rushing past them, forced through the throat by contractions of the diaphragm. This variation in density is slight—only one part in ten thousand on average distinguishes the compressed from rarefied adjacent regions. The larger the density variation, the greater the volume or loudness of the sound wave. You only have control over the initial density variation as the sound leaves your mouth. The compressed region of air expands, compressing the region in front of it, which in turn expands and squeezes the next region of air. What you hear is the instruction set for the sound wave, generated from your vocal cords and transmitted to your ears. The air from your mouth does not physically travel from you to the listener. If I tell you that I had garlic for lunch, you hear this information before you come close enough to receive independent confirmation of this fact. As the information is spreading out in all directions, a large distance away from the speaker the variation in air density—the sound wave—is attenuated in magnitude, and will be below the threshold of detection if one is too far away.

Alternatively, if one is too close to the source, the eardrum is unable to linearly respond to the density variations, and the ability to distinguish differing sounds is degraded. This can come in handy, as the DC Comics miniaturized hero the Atom discovers in "The Case of the Innocent Thief" in *Atom* # 4.* In this story, a crook named Elkins discovers a hypnotic ray that compels anyone to obey any oral commands he hears. While the Atom is only a few inches tall, the crook exposes him to this ray and yells a command forbidding the Atom to try to capture him. Yet almost immediately the Atom knocks out Elkins, using a pink rubber eraser as a trampoline to come within punching distance. The Atom is able to resist the mesmeric command because, as he explains at story's end, "in his excitement, Elkins shouted his orders at me—words which sounded like thunder to me! And since I couldn't understand a word he said to me, I didn't have to obey him!" Interestingly enough, my kids put forth this same argument nearly daily, though they have not yet mastered miniaturization technology, nor am I shouting.

An additional difficulty that accompanies shrinking to the size of an ant is that Ant-Man's vision will be blurred. The average spacing between adjacent peaks or valleys in the alternating electric and magnetic fields that comprise a light wave—called the "wavelength"—determines the color of the light. Let us say that on average white light—which consists of light of all wavelengths from red (650 nanometers) to violet (400 nanometers) added together in equal magnitudes—has a wavelength of 500 nanometers (one nanometer is one billionth of a meter). In order for light to be detected, it must be incident on the rods and cones in the back of your eye, and in order to get to these photo-receptors it must first pass through your pupil. This opening in the front of your eye, depending on the brightness in the room in which you are reading this, is roughly 5 millimeters in diameter. One millimeter is equal to a million nanometers, so the opening in your pupil is roughly 10,000 times larger than the wavelength of visible white light. From the point of view of the light waves, the pupil is a very large tunnel through which they can easily pass. When Ant-Man shrinks to the size of an insect, however, the opening in his pupil will be 300 times

---

* We'll have more to say about the Atom and his shrinking ability in chapter 12.

smaller than at his normal height. The orifice in his eye is now roughly a factor of 30 times larger than the wavelength of visible light, which is still 500 nanometers. The light waves can still fit into this "tunnel," but only just.

To understand what the consequences are when the size of the opening is only a few times bigger than the wavelength of light, consider water waves on the surface of a large lake. A channel is formed between two docks that ride low on the surface of the water. When the separation of the docks is very large, say nearly half a mile apart, compared with the spacing between peaks of the water waves, the waves pass through this region with no noticeable perturbation. Right near the dock, as the waves break, there is a change in the wave front, but in the middle between the docks the waves are hardly affected by the docks. This is the situation for Henry Pym at his normal height, when his pupil is 10,000 times larger than the wavelength of light. For the miniaturized Ant-Man, it is as if the two docks narrow to a bottleneck, so the separation of the docks is only a few times more than the separation between adjacent wave peaks. The waves still move through the constriction, but as they scatter off the edges of each dock they set up a complicated interference pattern on the other side of the obstruction. This effect is termed "diffraction" and is most noticeable when the dimensions of the object scattering a wave are comparable to the wavelength. If you were hoping to gain information about the cause of the water ripples by examining the wave fronts, you would obtain a clean, sharp image when the docks were thousands of feet apart and a distorted and confusing picture when the docks are only separated by a few feet.

The effect for Ant-Man is that the image he observes through his shrunken pupil will be blurry and out of focus. This is why an insect's eye, and in particular its lens, is radically different from the lens in a human's or larger animal's eye. Insects use compound lenses that adjust for the diffraction effects. Even so, it would be hard for a fly to read a newspaper, even if he cared about current events. An insect's eye is very good at detecting changes in light sources (such as the moving shadow created by an impending rolled up newspaper of doom), but poor at discerning the contrast between sharp edges. Consequently they rely on other senses, such as smell and touch (hair filaments detect subtle variations in air currents) to navigate their way through the wide world. Unfortunately for Ant-Man, the one sense that is least affected by miniaturization, smell, is the one that is least sensitive in humans.

DOES SIZE MATTER?–

THE CUBE-SQUARE LAW

THERE ARE ONLY SO MANY adventures a quarter-inch-tall character can have before the novelty of a super-small superhero wears off for the reader. A hint that the bloom may have left the Ant-Man rose can be found in *Tales to Astonish* # 48, wherein the supervillain the Porcupine captures Ant-Man and fiendishly attempts to eliminate him by placing him in a bathtub partially filled with water (as shown in fig. 14). Given that he cannot climb up the slick porcelain walls, Ant-Man is forced to endlessly tread water until he becomes exhausted, at which point he will drown. While the peril for Ant-Man is real, it's hard to get too worked up about a death trap consisting of a partially filled bathtub. Even the caliber of villains Ant-Man faced at this stage reflected the difficulty in keeping this concept fresh. The Porcupine was Alex Gentry, an engineer who uses his technological prowess to develop a quill-coated suit that conceals a host of offensive and defensive weapons—such as tear gas, stun pellets, ammonia (presumably for clean escapes), "liquid fire" (I suspect he means a gas flamethrower), detector mines, liquid cement, and others—with which he embarks on an obligatory crime wave. Now, I must say that, as a physics professor, I have worked with many engineers in my academic career, and it has been my experience that very few of them dress like giant porcupines.

If one of the functions of comic books is wish-fulfillment for their young readers, then it must be recognized that not many kids fantasize about how cool it would be to be only a few millimeters

Fig. 14. The cover of Tales to Astonish # 48, in which the Ant-Man is helpless in the fiendish death trap of a partially filled bathtub, at the mercy of engineer Alex Gentry, also known as the Porcupine (despite the boast on the cover—the Porcupine was actually fairly easy to forget)!

high or dress like a porcupine. Now, to be able to grow ten feet tall, that would be something. And so, in the very next issue of *Tales to Astonish* # 49, bowing to the inevitable, Henry Pym discovered a reverse version of his shrinking potion that enabled him to grow larger than his normal height of six feet. And so Giant-Man was born into the Marvel universe. Eventually Pym became a diminutive crime-fighter again, this time as the flamboyant Yellowjacket, but for the majority of his superhero career Pym fought for justice as either the oversize Giant-Man or Goliath (same hero and same power of super-growth, just different code name and costume).

And yet, it turns out that being larger than normal carries with it a different, yet no less pressing, set of physical challenges. For one thing, as pointed out in *Ultimates* # 3 (a modern version of the Avengers, featuring a new Giant-Man), your now much-larger (super-dilated) pupils would let much more light into your eyes. You would therefore always need to wear special goggles while enlarged, to avoid overloading your optic nerves. In addition, there is a fundamental limit to the size that someone could grow; assuming of course that one could grow far beyond his normal height in the first place, which requires just as much of a "miracle exception" as does miniaturization. This limitation is set by the strength of materials (particularly bone) and gravity. Gravity enters the situation because your mass will increase in proportion to your volume if you maintain a constant density. Density is mass divided by volume, so the bigger you are (that is, the larger your volume), the greater your mass if the ratio of the two (density) remains unchanged. You would be a less-than-imposing figure if your super-growth kept your mass fixed. In this case, the larger you became, the smaller your density would be.

Just such a situation faced Reed Richards of the Fantastic Four, when he encountered the monstrous alien invader Gormuu in *Fantastic Four* # 271. This issue relates a story that took place before Richards and his three comrades had been exposed to the cosmic rays that would give them their superpowers. Gormuu was a warrior invader from the planet Kraalo, a twenty-foot-tall misshapen green creature. This story was a nostalgic tip of the hat by writer and artist John Byrne to the Monster Invader from Space stories that dominated Marvel comics in the late 1950s until the Fantastic Four arrived to save both the universe and the financial fortunes of Stan Lee and Jack Kirby. The creatures in these comics (*Tales to Astonish, Strange Tales, Journey into Mystery,* and *Tales of Suspense,* before the Silver Age renaissance of superheroes) were always at least as big as a house and all had names with adjacent double letters, such as Orrgo (the Unconquerable), Bruttu, Googam (son of Goom), and Fin Fang Foom. Gormuu's competitive advantage regarding world conquest was that his size increased whenever he was struck with any form of broadcast energy. Richards, upon examining a ten-foot-long and inches-deep footprint left behind by the already-enormous alien, realizes that the only way to stop this menace is to continue to strike him with more and more broadcast power. For any creature large enough to leave a ten-foot-long footprint behind should also leave an impression several feet deep, *if* its mass were increasing at the same rate as its volume. Discovering that Gormuu's growth was at constant mass, *not* constant density, Richards feeds the alien so much energy that he grows larger than the Earth, and less dense than the background of space, until he becomes an insubstantial and decidedly unthreatening footnote in our nation's atomic-age history. Bearing in mind the cautionary tale of Gormuu, let us assume that Henry Pym, in his guise as Giant-Man, manages to maintain a constant density as he grows, so that his weight increases in uniform proportion to his volume. In order to treat the Giant-Man situation mathematically, we will have to make a simplifying assumption—namely, that Henry Pym is a giant box.

When dealing with the messy complications of the real world, physicists sometimes have to make drastic simplifications in order to make any progress in understanding nature. This tendency toward reducing a problem to its most basic components is reflected in the tale of the chicken farmers and the theoretical physicist: It seems that a group of chicken farmers had acquired an assortment of prize-winning hens at the State Fair. When they brought these fowl to their henhouse, they discovered that there was one problem with these poultry: no eggs. The farmers tried everything they could think of to get the chickens to begin laying eggs: mood music piped into the henhouse, track lighting, shag carpeting, but all with the same result—no eggs. In desperation they called in a theoretical physicist to give them some advice. After studying the problem for a week from all points of view, the scientist called the farmers together in a lecture room and triumphantly announced that he had solved the problem. However, the farmers' spirits fell when the physicist began his lecture by drawing a large circle on the blackboard and stating: "Assume a spherical chicken . . ." But sometimes this approach, beginning with a spherical chicken for example, is the right one. When viewed from a very great distance, after all, chickens do indeed appear spherical—at least, if you squint enough. If the spherical model captures some essential aspect of the chicken problem, then one can later add lumps to build a more accurate description of the hen. On the other hand, it is possible that the spherical starting point is too simplistic and misses out on the basic physics underlying the problem. The alternative, however, is to drown in a sea of technical details, some being crucial while others are irrelevant to the problem at hand. Deciding what factors must be included in the initial spherical model and what can be safely deferred until later usually only comes with practice.

So, perhaps you'll forgive me when, in discussing Giant-Man, I assume a cubical Henry Pym. For the argument I intend to make, the math is much simpler if Giant-Man is taken to be a large box with equal sides of length l. Of course, he is more naturally described as a large cylinder, but I want to keep the math as simple as possible. Well, if he's a box, then his volume is the product of his length, height, and width. If each side of the cube has a length l, then his volume is $l \times l \times l$. A box that has a length of ten feet, a width of ten feet, and a height of ten feet will have a volume of 10 feet $\times$ 10 feet $\times$ 10 feet, or 1,000 feet cubed. This unit of volume is sometimes called "cubic-feet" or feet^3, which represents the fact that we multiplied a length (feet) times another length times another length. Now assume that Giant-Man uses his Pym particles to grow twice as large in all directions. So his height is now 20 feet, and his width and length are also 20 feet each. In this case his volume is 20 feet $\times$ 20 feet $\times$ 20 feet—that is, 8,000 cubic-feet. So, doubling his length in all three directions increases his volume by a factor of eight. If his length increased in all three directions by a factor of ten, such that his length, width, and height were now 100 feet as opposed to 10 feet, then his volume would be 1,000,000 cubic-feet—one thousand times bigger than his initial volume of 1,000 cubic-feet.

If Giant-Man is going to maintain a constant density as he grows, then his mass must increase at the same rate as does his volume, not his length! Doubling his height (as well as his width and breadth) requires that his weight also go up by a factor of eight to keep his density the same. Well, Hank Pym gets heavier the bigger he becomes—so what? The problem is that his weight increases faster than does the ability of his skeleton to carry his weight, so at a certain height Giant-Man risks breaking his legs simply by standing up. The strength of an object, or its resistance to bending or to being pulled apart by a pull or crushed by a push, depends on how wide it is and not on its length. The technical way to say this is that an object's "tensile strength" is determined by its cross-sectional area.

Think of a fishing line rated for 20 pounds—that is, for holding up a 20-pound fish. A heavier fish would snap the line when we tried to lift it onto the boat. If we want to keep a heavier fish suspended, changing the length of the fishing line will be absolutely no help whatsoever. To increase the strength of a fishing line, don't increase its length but rather its diameter.* The wider the fishing line, the greater the area over which the force pulling it is distributed, and the lesser the force applied across any given tiny element holding the line together. When a fishing line, or anything for that matter, breaks, the chemical bonds that were holding the material together snap and fly apart. The larger the area that's available to support a given force, the less stress or strain is applied to any particular molecule, and the less likely it is that a catastrophic failure will occur. When breaks do happen, it is usually because a molecular imperfection or flaw magnifies the applied force locally, leaving the material weaker than if it were uniform and atomically perfect.

The dependence of a material's strength on its cross-sectional area limits the distances over which Mr. Fantastic of the Fantastic Four can stretch. After being bombarded with cosmic rays during the ill-fated maiden voyage of the spaceship he designed, Reed Richards gained the ability to stretch or compress any part of his body. But as explained in *Fantastic Four Annual* # 1, he cannot extend the length of his body beyond 500 yards. A wooden two-by-four board (that has a rectangular cross-section of two inches by four inches) that is three feet long can be supported above the ground by a sawhorse at either end and remain parallel to the floor. A six-foot-long board of the same cross-sectional area would sag slightly in the middle, if still supported at the ends. A sixty-foot-long board would droop considerably, while a six-mile-long board would touch the ground at its center even ignoring the curvature of the Earth. As Reed points out in *Fantastic Four Annual* # 1, "the further my body stretches, the weaker my muscles

---

* Apple trees employ this principle as part of their seed-dispersal mechanism. When the fruit has matured and reached a sufficient mass, its weight exceeds the tensile strength of the narrow stem by which it is attached to the tree. When the stem snaps, the fruit consequently falls to the ground, where it is consumed by fauna, and its seeds are distributed to other locations.

become,* so that I cannot exert as much force stretched to a great distance as I can extending for a shorter distance." Reed Richards, whose understanding of the mass-volume relationship saved the Earth from the terror of Gormuu, is also a walking illustration of the cube-square law of tensile strength.

As our cubical Henry Pym grows into Giant-Man, his volume increases faster than his cross-sectional area. The compressive strength of an object, such as the femur in your thigh or the vertebrae in your spine, is determined by its cross-sectional area—that is, the area of one of its faces if it were a rectangular solid. As Giant-Man grows larger, his bones naturally increase in size proportionally with the rest of his body. The femur's or vertebra's strength grows by the square of his expansion rate. But the bigger that Giant-Man becomes, the more weight his bones must support—at constant density his weight increases by the cube of his growth factor. Suppose at his normal height Dr. Pym is six feet tall and weighs 185 pounds. His femur at his normal height can support a weight of 18,000 pounds while a single vertebra can support 800 pounds—indicating that nature builds considerable redundancy into some crucial load-bearing structural components. The femurs of elephants and dinosaurs are thicker and denser than those of humans, while mice and birds have correspondingly thinner and lighter bones. At a height of sixty feet, Giant-Man's expansion factor is ten. His volume therefore grows by a factor of 1,000 while the cross-sectional area of his bones only increases by a factor of 100. Henry Pym would now weigh 185,000 pounds, while his vertebrae could only support a weight of 80,000 pounds and his femur could carry 1,800,000 pounds. At this height, his skeleton is not able to uniformly support his weight.

In order to keep Giant-Man from growing so tall that no villain, even a superpowered one, could pose a credible threat to him, Stan Lee argued in the 1960s that increasing his size placed a biological strain on Henry Pym. His optimal strength was at a height of roughly 12 feet, and if he grew to more than 40 or 50 feet, he was as tall as a house but as weak as a kitten. Years later, the metabolic limitation was replaced by a physical one derived from the cube-square law. As illustrated in fig. 15, from *Ultimates* # 2, it is now

---

* Offset somewhat by the increase in torque by the longer moment arm (see chapter 8).

Fig. 15. *A scene from* Ultimates *# 2, a 2002 updating of the Avengers. Here Henry Pym and his wife, Janet van Dyne (aka the Wasp), prepare for Pym's first experimental test of his growth serum. Dr. Pym worries that if he exceeds a height of sixty feet, his thigh bones will be unable to support his body mass.*

recognized that even if the metabolic issues involving a growth serum are resolved, gravity and physics will still provide stringent limits on your ultimate size (though as we've just seen, his back would break before his thigh bones snapped).

The fact that volume grows faster than surface area is true even for non-cubical objects. The volume of a sphere is given by the mathematical expression of a constant $(4\pi/3)$ times the radius of the sphere cubed, or $(4\pi/3)\,r^3$, while its surface area is 4π times the radius squared, or $4\pi\,r^2$. (A volume will always have the units of a length cubed—such as cubic-feet—while an area has the dimensions of a length squared, such that carpeting is measured in square yards.) Consequently the rising bubbles in a vat of acid into which Batman and Robin are being slowly lowered provides the Caped Crusader with a textbook illustration of the physical principle of the cube-square law.

If you've ever thought that the bubbles in your champagne or beer glass rose faster as they neared the top of your fluted crystal, it wasn't the alcohol affecting your judgment. The bubbly drink is supersaturated with carbon dioxide (the same gas that gives soda pop its fizz and beer its foam), which means that the pressure of the carbon dioxide gas forced into the liquid is greater than atmospheric pressure. When the cork is popped on the champagne bottle, or the twist-off cap is released on a bottle of soda or beer (or, if you're really cheap, when the twist-off cap is removed on your champagne bottle), there is a popping or hissing noise, due to some of the excess gas under pressure rapidly exiting the container. There is still additional carbon dioxide in the liquid, which collects into bubbles near small imperfections in the glass and then, being lighter than the surrounding fluid, rises to the top. The buoyancy force lifting the bubble is directly related to its spherical volume, which depends on the cube of the bubble's radius. The drag resistance force slowing down the bubble depends on its surface area (the bigger the area, the more champagne has to be pushed out of the way of the rising bubble) that in turn grows by the square of the bubble's radius. As the bubble moves through the champagne, it sweeps up additional carbon dioxide molecules dispersed throughout the liquid, becoming bigger in the process. There is thus a net excess force raising the bubble. If there's a force, there is an acceleration (Newton's second law holds even inside a glass of champagne) and the bubble will rise faster and faster.

If I had an infinitely tall glass, would the bubble accelerate to the speed of light? No; in chapter 5 we noted that the drag resistance depends not only on the surface area but also on the speed of an object (it takes more effort to push the fluid out of the way of a rapidly moving object than one progressing slowly). As the bubble rises and moves faster and faster, an additional drag force eventually balances out the upward lift, and once there is no net force, the bubble continues to move at a steady uniform speed (Newton's first law) known as the terminal (or final) velocity.

Now would be a good time for you to pause for some hands-on experimentation with champagne, beer, or soda to verify some of what you have just learned—but purely in the interest of science, of course!

SECTION
2

ENERGY— HEAT AND LIGHT

11

THE CENTRAL CITY DIET PLAN—

CONSERVATION OF ENERGY

THE FLASH MAY BE ABLE to run across the ocean and pluck bullets from the air, but a more important question concerns us: How frequently does he need to eat?

The short answer is, a lot! A more basic question we could ask is: *Why* does he need to eat? What exactly does food contain that is essential for any activity, whether running, walking, or even sitting still? And why do we only obtain these qualities from organic matter, and not from rocks or metal or plastic?

The Flash eats for the same reason we all do: to provide raw materials for cell growth and regeneration, and to provide energy for metabolic functioning. At birth, your body contains a certain quantity of atoms that was insufficient to accommodate all of the growth that occurred during your lifetime. As you grew and matured, you needed more atoms, typically provided in the form of complex molecules that your body would break down and convert into the building blocks necessary for cell replacement and growth. As noted in our discussion of the explosion of Krypton, all of the atoms in the universe—including in the food we ingest—were synthesized via nuclear reactions in a now long-dead star where hydrogen atoms were squeezed together to form helium atoms, helium fused to form carbon, and so on. An additional by-product of these fusion reactions in our sun provides the second essential component of the food we eat. Matter-Eater Lad of the Legion of Super-Heroes may be able to subsist by consuming inert objects such as metal or stone, and the cosmic menace Galactus must consume the life-energy of planets, but for the most part the food we eat must have been previously alive. Only such foodstuffs provide us with an additional necessary component, as mysterious as its name is mundane: Energy.

The use of the word "energy" is so common that it is unnerving to realize how difficult it is to define without using the word "energy" or "work" in the explanation. The simplest nonmathematical definition is that "energy" is a measure of the ability to cause motion. If an object is already moving, we say it possesses "kinetic energy," and it can cause motion if it collides with something else. Even if it is not moving, an object can possess energy, such as when pulled by an external force (for example, gravity) but it is restrained from accelerating (say, by being physically held above the ground). Since the object will move once it is let go, it is said to possess "potential energy."

All energy is either kinetic or potential, though depending on the circumstances a mass can have both kinetic and potential energy, such as when Gwen Stacy fell from the top of the bridge in chapter 3. When she was on the top of the bridge, she had a large potential energy, as gravity had the potential to act on her over a long distance. But her motion was constrained, as the bridge was holding her up. When she was knocked off the top of the tower, the constraint was removed, and the force acting on her (gravity) then began her acceleration. As she plummeted, she had a shorter and shorter distance over which to continue falling, so her potential energy decreased. The potential energy didn't disappear, but rather her large potential energy at the top of the bridge was converted to ever-increasing kinetic energy as she fell faster and faster. At any given point in her fall, the amount of kinetic energy she gained was exactly equal to the amount of potential energy she lost (ignoring the energy expended in overcoming air resistance). If she were to have struck the water at the base of the bridge, her potential energy would have been a minimum (once at the base of the tower she has no more potential to fall), while her speed and hence kinetic energy would have been at its largest. In fact (again ignoring air drag) her kinetic energy at the base would have been exactly equal to her large potential energy at the top of the bridge when she started to fall. This kinetic energy is then transferred to the water, which supplies a large force that changes her high velocity to zero, with the same dire results as when she was caught in Spider-Man's webbing as described in chapter 3.

Or imagine Spider-Man swinging back and forth on his webbing, like a pendulum. At the top of his arc he is not moving (which, again, is why it's the highest point of his swing), but he is high off the ground, and has a large potential energy. His potential energy at the bottom of his swing is a minimum, and if he had started at this point, he would not move. Starting at a higher point, his earlier potential energy changes to kinetic energy, and at the lowest point in his arc, his loss of potential energy is exactly equal to his gain in kinetic energy. The only force acting upon him at this lowest point is gravity (straight down) and the tension in the webbing (straight up). Neither of them acts in the horizontal direction of his swing at this point. But he is already moving, and an object in motion will remain in motion, unless acted upon by an external force. As he overshoots this lowest point and starts to rise again, his kinetic energy changes back to potential energy. If no one pushes him, he can never have more total energy than what he started off with (where would it come from?), and so the final point of his swing cannot be higher than his initial starting height. In fact, some of his kinetic energy is used up in pushing the air out of his way (air drag), so he will rise to a lower height than his starting point.

This accounting of how much energy is potential and how much is kinetic implies one of the most profound ideas in all of physics: *Energy can be neither created nor destroyed, but can only be transformed from one form to another.* This concept goes by the fancy title of the **principle of conservation of energy**. We have never been able to catch Nature in a slipup where the energy at the start of a process does not exactly equal the energy at the end. Never.

When physicists studied the decay of radioactive nuclei in the 1920s and 1930s, they found that the final energies of the emitted electrons and resultant nuclei did not equal the starting energy of the initial nuclei. Faced with the possibility that energy was not being conserved in the decay reaction, Wolfgang Pauli instead suggested that the missing energy was being carried away by a mysterious ghost particle that was invisible to their detectors. Eventually devices were constructed to observe these "ghost particles." They not only turned out to be real, but in fact neutrinos (as these mystery particles were named, somewhat whimsically, by Enrico Fermi, describing them in Italian as "little neutral ones") are some of the most prevalent forms of matter in the universe.

Consider driving a nail into a wooden board. The potential energy of the hammer, held over a carpenter's head, is converted to kinetic energy as she swings. When the hammer strikes the nail, the hammer's kinetic energy causes the motion of the nail (hopefully deep into the board of wood) and, as a side effect, also causes the atoms in the head of the nail to shake more violently, warming up the nail. The partitioning of the incident kinetic energy of the hammer into additional vibrations of the atoms in the nail head, the forward motion of the nail itself, and the breaking of molecular bonds in the wood (necessary if the nail is to occupy space previously claimed by the board) can be summarized by describing the "efficiency" of the hammering process. If one carefully adds up all of the small and large bits of kinetic energy in the nail, wood, and even the air (the "bang" one hears upon striking the nail results from a pressure wave—sound—induced in the surrounding atmosphere), the net result will exactly equal the initial kinetic energy of the hammer right before it strikes the nail head. However, warming the nail and creating a sound effect are "waste energies" from the point of view of the carpenter, and count against the efficiency of the hammering process.

Sometimes this waste energy is not so insignificant. An automobile traveling along a level road has a kinetic energy. This energy arises from a chemical reaction during the combustion of gasoline vapor with oxygen, ignited by an electrical jolt from the sparkplug. The gases resulting from this small explosive reaction are moving at great speeds, so that they may displace a piston. The piston's up/down motion is translated via an ingenious system into the rotation of the car's tires. Of course not all the energy of this chemical reaction goes into the displacement of the auto's pistons—much of it heats up the engine, which is useless from the point of view of locomotion. In addition, as the car travels down the highway, energy is also needed to push the air out of the way of the car. An automobile's efficiency is determined in large part by the effort of displacing air from the immediate volume it intends to occupy— over six tons of air for every mile traveled by an average-size car! The larger the profile of a car or truck, the greater the volume of air that must be displaced, and the more energy must be devoted to this task, in addition to propelling the car forward. This same principle also explains why it is easier to run underwater through a swimming pool with your hands flat at your sides than if you hold them out away from your body. The smaller the surface area, the greater the fuel efficiency for comparable mass vehicles. The aerodynamics of a sports car's design is therefore not only intended to increase our attractiveness to the opposite sex, but is also paramount in determining how frequently we must visit a filling station.

The Green Goblin expended energy carrying Gwen Stacy to the top of the George Washington Bridge. This increase in her potential energy was stored as she lay atop one of the towers. The increase in her potential energy came from the chemical energy in the fuel in the Goblin's glider, and so on. Taken to its logical conclusion, the Principle of Conservation of Energy states that, if one can never create new energy or destroy current energy, but simply convert it from one form to another, then all of the energy currently in the universe was present at the moment of the Big Bang that heralded the universe's creation. At this primordial instant, the entire universe was compressed within a volume smaller than a single electron. There was no matter at this point, only energy, squeezed into an inconceivably small volume. As the universe expanded, the amount of energy remained unchanged but was now spread over an ever-increasing volume.

"Energy density" is the energy per volume; consequently, if the amount of energy remains the same but the volume increases, the energy density will decrease. All of the matter present in the universe today came into existence through a process represented by Einstein's famous equation $\mathbf{E = mc^2}$ when the energy density lowered to a critical point. The import of $\mathbf{E = mc^2}$ is that matter can be considered "energy slowed down." As the universe expanded and cooled down, within the first second after the Big Bang, the energy density was low enough that matter, such as protons and electrons, started condensing into existence, not unlike cooling water forming ice crystals as the temperature is lowered below the freezing point. This spontaneous formation of matter only occurred once—earlier in the universe's history the energy density was too high to allow protons and electrons to condense into existence, while at later times (such as now), when the energy density is below the $\mathbf{E = mc^2}$ threshold, there is not enough background energy in outer space to spontaneously form matter.* The protons and electrons created in the early eons of the universe came together due to their electrostatic attraction and formed hydrogen atoms. Gravity pulled some of these hydrogen atoms together to form large clumps that became stars. In the centers of these stars, held together by gravitational potential energy, a nuclear reaction transforms these hydrogen atoms into heavier elements and kinetic energy.

Now, it is all well and good to say that all of the energy (and consequently all of the matter) found in the universe today was present at the moment of the Big Bang. But this only leads to two deeper questions about energy: What is it *really*? And where did it originally come from? Science provides the exact same answer to both questions: Nobody knows.

FAST FOOD

In order to figure out how much the Flash must eat to be able to run at super-speed, we need to calculate his kinetic energy. Physicists are always looking to conserve energy—consequently, we'll recycle the math from chapter 1 so we won't have to do any more work. Speaking of work, in order to change the kinetic energy of an object, by either speeding it up or slowing it down, one must do Work. "Work" is capitalized, because in physics the term has a specific meaning that is slightly different from its common usage.

When a force acts on an object over a given distance, we say that the force does **Work** on the object and, depending on the force's direction, will either increase or decrease the object's kinetic energy. In this way Work is just another term for energy, and they will have the same units. For a falling mass **m**, the force acting on it is its weight due to gravity $F = mg$, and the distance the force acts upon the object is just the height **h** that it falls. So **Work = (Force) × (distance) = (mg) × (h) = mgh**. This turns out to be

---

* In 1997 scientists were able to directly verify that a sufficiently large energy density could cause matter to spontaneously come into existence. By colliding high-energy gamma-ray photons of light together, they were able to create electron/antielectron particle pairs in the laboratory, in essence recreating the mechanisms operating in the first seconds of the universe.

the potential energy that the object had at a height **h**, so in this example Work can be viewed as the energy needed to increase an object's potential energy.

Consider the falling Gwen Stacy from chapter 3 or the leaping Superman from chapter 1. In either case the Work that gravity does is given by **Work = mgh**. For Gwen the Work increases her kinetic energy, and for Superman it decreases his kinetic energy. The difference is, for Gwen the force pulls her down in the direction of her motion, while for the Man of Steel the force is still downward, but is opposite to the direction of his leap. Gwen starts with no kinetic energy, but the gravitational force acting over a distance (the height of the bridge tower) provides her with quite a large final velocity right before striking the water. The connection between her final speed **v** and the distance she fell **h** was given by $v^2 = 2gh$, where **g** is the acceleration due to gravity. This is a true statement, and according to our Rule of Algebra (see Preface), we can multiply and divide both sides of a true statement by the same quantity, and it remains a true statement. So if both the left- and right-hand sides of $v^2 = 2gh$ are divided by 2, the result is $v^2/2 = gh$. If we now multiply both the left- and right-hand sides by Gwen's mass **m**, we obtain $\frac{1}{2} mv^2 = mgh$. The right-hand side is the **Work** that gravity does on Gwen. The left-hand side must therefore describe her change in kinetic energy—that is, her final kinetic energy minus her initial kinetic energy. Since she started with no kinetic energy (no motion, no kinetic energy, though she had plenty of potential energy), her final kinetic energy is stated as **Kinetic Energy** $= \frac{1}{2} mv^2$. Congratulations—you've just done *another* physics calculation.*

* * *

When the Flash stops running, Work is done changing the Scarlet Speedster's kinetic energy. From time to time, the acceleration or deceleration that the Flash must experience is (more or less) real-

---

* It turns out that we could have started our discussion in chapter 1 with a definition of energy and, using the Rule of Algebra, worked backward to "derive" the expression **F = ma**, instead of starting with **F = ma** and determining the energy as we did. Where one starts in the calculation is a matter of personal taste. In the end we will always find the expression for Gwen Stacy of $v^2 = 2gh$. This connection between her final velocity and the height from which she falls is the important thing, and Gwen does not really care which equations one uses to obtain it.

Fig. 16. *A rare example from* Flash *# 106 of the realistic effects of the Flash's sudden deceleration. The shorter the stopping distance, the greater the force his boots must exert on the ground when braking.*

istically addressed, and the consequences of these decelerations portrayed. In *Flash* # 106, the Flash needed to stop suddenly while chasing an object that was traveling at 500 mph. The comic shows him gouging giant ruts in the ground with his feet as he attempts to quickly bring himself to rest. Here the forces, friction in particular, that would accompany his rapid deceleration are accurately represented. In bringing himself to rest from a speed of 500 mph, the large change in kinetic energy requires a correspondingly large **Work**. The comic panel (fig. 16) shows the Flash stopping in approximately fifteen feet, so the distance is short, and since **Work = (Force) × (distance)**, the force that his feet exert on the ground must be correspondingly very large. In fact, to change his velocity of 500 mph in a distance of fifteen feet requires a force of over 80,000 pounds!

Similarly, in *Flash* comics vol. 2, issue # 25 (April 1989), Wally West* runs so fast that, in his attempt to stop suddenly, he leaves mile-long gashes across North America. From the length of the skid marks, the scientists who are tracking Wally are able to determine both how fast he was going and his probable stopping point, using

---

* Wally was originally Kid Flash but by 1985 he had dropped the "kid" portion of his title, taking over the mantle of the Flash after Barry Allen, the Silver Age Flash, had died saving our universe from the Anti-Monitor. It's a long story.

the same techniques that the police employ when reconstructing an automobile accident from the length of tire skid marks. Realistically, one should *always* know where the Flash has been from the deep gouges his feet would excavate every time he suddenly started or stopped running. Fortunately for the Central City Department of Roads and Transportation, this physically accurate portrayal of the Flash's powers only occurred occasionally.

Returning now to the Flash's eating habits, if kinetic energy **KE** is written mathematically as $\mathbf{KE = (1/2)\ mv^2}$, then the Flash's caloric intake requirements increase quadratically the faster he runs. If he runs twice as fast, his kinetic energy increases by a factor of four, and thus he needs to eat four times more in order to achieve this higher speed. Back in the Silver Age (late 1950s to 1960s), artist Carmine Infantino would draw Barry Allen as fairly slender and not as a hulking mass of muscle, since he was, after all, a runner (the Flash, that is, not Carmine). If the Flash weighed 155 pounds on Earth, then his mass would be 70 kilograms. When running at 1 percent of the speed of light (nowhere near the Flash's top speed), his speed would be $\mathbf{v} = 1,860$ miles/second or 3 million meters/second. In this case his Kinetic Energy **KE** is $(1/2) \times (70\,\text{kg}) \times (3,000,000\ \text{meters/sec})^2 = 315$ trillion kg-meter^2/sec$^2 = 75$ trillion calories. Energy is so frequently used in physics that it has its own unit of measurement, one of which is termed the "calorie," and is defined as 0.24 calories $= 1$ kg-meter^2/sec^2. That is, 0.24 calories is equal to the Work resulting from applying a force of 1 kg-meter/sec^2 over a distance of one meter.

The reason that 1 kg-meter^2/sec^2 is equal to this odd number of calories (0.24, to be exact) is due to the fact that in the mid nineteenth century, physicists were confused about energy, a situation that has not greatly improved in the intervening years. A calorie was originally defined as a unit of heat, as heat was thought to be a separate quantity distinct from Work and energy. Hence one system of measurement for heat was developed, while a different unit was employed to measure kinetic and potential energy. The physicist who recognized that heat was simply another form of energy, and that mechanical work could be directly transformed into heat, was James Prescott Joule, in whose honor a standard unit of energy (1 Joule $= 1$ kg-meter^2/sec^2) is named. While physicists use Joules when quantifying kinetic or potential energy, we'll stick with the more cumbersome kg-meter^2/sec^2 in order to emphasize the different factors that enter into any determination of energy.*

It should be noted that a physicist's calorie is not the same as a nutritionist's calorie. To a physicist, one calorie is *defined* as the amount of energy needed to raise the temperature of one gram of water by one degree Celsius. This is a perfectly valid, if arbitrary, way of defining energy in a laboratory setting. But this definition leads to the observation that a single soda cracker contains enough energy to raise the temperature of 24,000 grams of water by one degree. That is, to a physicist the energy content of one plain cracker is 24,000 calories! In order to avoid always dealing with these very large numbers, a Food Calorie is defined to be equal to 1,000 of these "physics calories." Therefore the 24 Food Calories in a single cracker are actually equivalent to 24,000 calories using the physics laboratory definition of the term. Just as well, as it's bad enough to think of the roughly 500 Food Calories in a cheese-burger, but if we considered that it actually contained 500,000 physics calories, we might never eat anything ever again!

To convert the Flash's kinetic energy of 75 trillion calories into Food Calories, we should divide his energy by 1,000. This helps, but he still expends 75 billion Food Calories running at 1 percent the speed of light. Put another way, he would need to eat 150 million cheeseburgers in order to run this fast (assuming 100 percent of the food's energy is converted into kinetic energy).† If he stops, his kinetic energy goes to zero, and in order to run this fast again, he needs to eat another 150 million burgers. At one point in *Flash* comics (during the mid-1980s) it was briefly acknowledged that he would need to eat nearly constantly (even chewing at super-speed) in order to sustain his high velocities. In the Golden Age, the Silver Age, and now in the Modern Age, conservation of energy is conve-

---

* Note that kg-meter^2/sec^2 is also the unit for gravitational potential energy **PE = mgh**, when kg, meter/sec^2, and meters are used respectively for mass **m**, acceleration due to gravity **g**, and height **h**. This is reassuring, for if kinetic energy is equivalent to potential energy, they should be measured with the same units. It would indicate something very wrong with our analysis if kinetic energy had units of kg-meter^2/sec^2 and potential energy or Work had units of sec^3/kg or something equally ridiculous.

† As roughly half of our caloric intake goes toward maintaining metabolic functions, the Flash would likely need to eat twice the amount we have estimated.

niently ignored. Nowadays the Flash's kinetic energy is ascribed to his being able to tap into and extract velocity from the "Speed Force," which is a fancy way of saying: Relax, it's only a comic book.

CHEESEBURGERS AND H-BOMBS

The next question we ask is why does a cheeseburger, or any food, provide energy for the Flash? It's easy to identify kinetic energy when something is moving, and the potential energy due to gravity is also pretty straightforward to understand, but there are many other forms of energy that require some thought as to which category, potential or kinetic, they belong. The energy the Flash gains by eating is not due to the kinetic energy of atoms shaking in his food (a hot meal has the same number of calories as a cold one) but from the potential energy locked in the chemical bonds in his food. As energy can never be created or destroyed, but only transformed from one state to another, let us follow the chain backward, to see where the stored potential energy in a cheeseburger comes from.

In order to understand the potential energy stored in food, we have to consider some basic chemistry. When two atoms are brought close to each other, if the conditions are right, they will form a chemical bond, and a new unit, termed a "molecule" will be created. A molecule can be as small as two oxygen atoms linked together, becoming an oxygen molecule (O_2), or it can be as long and complex as the DNA that lies within every cell of your body. The question of whether two or more atoms will form a chemical bond, and the elucidation of these conditions, is the basis of all of chemistry. All atoms have a positively charged nucleus around which a swarm of electrons hover. The chemical properties of an element are determined by the number of electrons it possesses, and how they manage to balance their mutual repulsion (as they are all negatively charged) with their attraction toward the positively charged nucleus. When an atom is brought very close to another atom, the most likely locations of the electrons from the two atoms overlap and, depending on their detailed nature, there will be either an attractive or repulsive force between the two atoms. If the force is attractive, the electrons create a chemical bond and the atoms form a molecule. If the force is repulsive, then we say that the two atoms do not chemically react. In order to determine whether the force is attractive or not involves sophisticated quantum mechanical calculations. (We'll have much more to say about quantum mechanics in section three.) If the force is attractive, and one restrains the two atoms to keep them physically apart, then there is a potential energy between the atoms, since once this restraint is removed the two atoms form a molecule. In this way, we say that the two atoms, once chemically joined, are in a lower-energy state, just as a brick's gravitational potential energy is lower when placed on the ground. Work has to be done to lift the brick to a height h, just as energy has to be supplied to the molecule to break it apart into its constituent atoms.

We are *finally* (and I can almost hear you saying: *Thank goodness!*) in a position to answer the question of why the Flash needs to eat. Or rather why food provides the energy he needs to maintain his kinetic energy. When the Flash runs, he expends energy at the cellular level in order to expand and contract his leg muscles. This cellular energy in turn came from the breakfast that Barry Allen ate. From where did the energy in the food arise? From plants, either directly consumed or through an intermediate processing step (such as animal meat). This stored energy in food is simply potential energy on a molecular scale. Plants take several smaller molecular "building blocks" and process them, stacking them up into a subcellular "tower of blocks." This molecular tower of complex sugars, once constructed, is fairly stable. The process of lifting and arranging a group of blocks into a tall tower raises the blocks' potential energy (except for the bottom block).

Similarly, plants do Work when constructing these sugars from simpler molecules, raising the potential energy of the final, synthesized molecule. The potential energy remains locked within the sugars until the mitochondria within our cells constructs Adenosine Triphosphate or ATP, releasing the saved energy, just as the Work in building a tower of blocks is stored as the potential energy of the top blocks until the tower is knocked down, converting their potential energy into kinetic energy. The amount of energy released by the ATP in the Flash's leg-muscle cells is greater than the energy needed to "knock the complex sugar tower down," though the gain to the Flash is much less than the plant cell's effort in raising the tower in the first place.

Where did the plant cell obtain this energy? Through a process termed "photosynthesis," whereby the energy in sunlight is absorbed by the plant cell and employed in complex sugar construction. The light comes from the sun (don't worry, we're nearly at the end of the line), where it is generated as a by-product of the nuclear fusion process, in which hydrogen nuclei are sintered together through gravitational pressure to create helium nuclei. Ultimately, all of the chemical energy in food is transformed sunlight that was generated by the nuclear fusion process that occurs in the detonation of a hydrogen bomb. In this way all energy on Earth is solar energy at its source, just as all the atoms on Earth, from the ATP molecules throughout the Flash's body to the ring in which he stores his costume, were created in a solar crucible (though obviously not that of our own sun).

Ultimately, all of life is possible because the mass of a helium nucleus (containing two protons and two neutrons) is slightly less than that of two deuterium nuclei (a deuterium nucleus contains one proton and one neutron) combined in the center of a star. And by slightly less I mean that the mass of a helium nucleus is 99.3 percent of two deuterium nuclei. This small mass difference leads to a large outpouring of energy, since from $E = mc^2$ the change in mass is multiplied by the speed of light squared.

That the mass of the resulting helium nucleus is exactly 99.3 percent of the reactants makes life as we know it in the universe possible. If the mass difference were 99.4 percent, then deuterium nuclei would not form, and hence fusion of helium could not proceed. In this case stars would shine too dimly to synthesize elements, and no violent supernova explosions would occur to both generate heavier elements and expel them into the void where they may form planets and people. On the other hand, if the mass difference were 99.2 percent, then too much energy would be given off from the fusion reaction. In this case protons would combine to form helium nuclei in the early universe, and no nuclear fuel would remain when stars formed. The source of this amazing fine-tuning of the basic properties of nature is the subject of current investigation.

DEEP-BREATHING EXERCISES

In order to run, the Flash needs the energy stored in food, which is locked up in complex molecules. We have described this energy as similar to the potential energy of a tower of blocks, that plants must do work on in order to stack. We transform this stored energy—after first consuming the plants—into kinetic energy when the tower is knocked over. But what is the trigger to topple this tower? How does the tower know when the cell needs the energy to be released? There's a lot of biochemistry that goes into the release of energy by the mitochondria in a body's cells, but the essential, rate-limiting step involves a chemical reaction of oxygen going in and carbon dioxide coming out. Without oxygen intake, the stored energy in the cell cannot be unlocked, and there's no point in eating. The faster the Flash runs, the more kinetic energy he manifests, the more potential energy stored in his cells he needs to release, and the more oxygen he needs to breathe. We've already discussed the fact that he would need to eat a staggering amount of food in order to account for the kinetic energy he routinely displays. What about his oxygen intake? Would the Flash use up all of the Earth's atmosphere as he ran?

To answer this question, we first need to know how much O_2 the Flash uses when he runs a mile. The volume of oxygen use by a runner will depend on his or her mass, and has been measured to be about 70 cubic centimeters of O_2 per kilogram of runner per minute, for elite athletes at a pace of a six-minute mile. Taking the Flash's mass to be 70 kg, he then uses nearly 30 liters of O_2 for every mile he runs (a liter is one thousand cubic centimeters). Let's assume that this rate of O_2 use remains the same even for much higher speeds. Thirty liters of O_2 contains under a trillion trillion oxygen molecules, and at a speed of 10 miles per second, this means that the Flash inhales about a trillion trillion O_2 molecules *every second*. That sounds like a lot, but fortunately there are many more O_2 molecules in our atmosphere than that. A lot more. In fact, very roughly, the Earth's atmosphere contains over ten million trillion trillion trillion O_2 molecules. So even at a rate of consumption of one trillion trillion molecules per second, he would have to run this fast (10 miles/sec) and breathe at this rate continuously for more than 500 billion years before he exhausted our oxygen supply.

The faster he ran, the quicker he would use up our air, but even running at nearly the speed of light (which he is capable of, but doesn't do very often) it would take him 27 million years of continuous running and breathing at this rate to exhaust the atmosphere. So, at least regarding this aspect of his super-speed, we can breathe easier.

The Earth's atmosphere may be safe, but of course this assumes that the Flash is able to breathe at all while he runs. That is, running at several hundred miles per hour, would he be able to even draw a deep breath? Fortunately for the Scarlet Speedster, he carries a reservoir of air with him whenever he runs. In *Flash # 167*, this region of stationary air (relative to the Flash) is described as his "aura," while in fluid mechanics it is termed the "no-slip zone." Whatever you call it, it's the reason golf balls have dimples.

To understand this, try this simple physics experiment at home: Turn on the cold water in your bathroom sink, but only barely open the valve. For the best results, first remove the aerator in the faucet. When the water is just barely coming out, you may see it move very smoothly from the faucet, with the appearance of a polished cylinder, wider at the faucet outlet and tapering slightly due to surface tension. Ignore the sound of the water splashing in the basin and it is difficult to tell that the water is moving at all, and isn't actually a rigid structure. This type of water flow, where all the water molecules are moving smoothly in the same direction, is termed "laminar flow." At the opposite extreme, turn the valve all the way open. The water churns and swirls out, moving in different directions and with a wide range of speeds. This type of water flow is termed "turbulent." Naturally, if you want to get water through a pipe in the most efficient manner possible, you would like the flow to be laminar, where all water molecules are moving in one direction down the pipe, rather than turbulent, where vortices and swirls by necessity mean that some water molecules are moving against the flow.

Even in laminar flow through a pipe, all the molecules may move in the same direction, but they won't all have the same speed. Those molecules at the outer edge will collide with the pipe's walls, transferring their kinetic energy to the pipe (which is rigid, so that the pipe warms up a bit but doesn't move) and coming to rest. Right next to the pipe's walls is a thin layer of water that is not moving. Water next to this non-moving layer loses some of its kinetic energy, but not all, because unlike the atoms in the pipe, the water molecules in this "no-slip zone" can move. In the next ring closer to the center of the pipe, the water is moving a bit faster still. So, even in uniform laminar flow, there are a continuous series of concentric rings, each ring moving progressively faster than the adjacent ring. The water dead center in the pipe moves the fastest of all. In laminar flow all rings are uniform, while in turbulent flow there is chaotic motion across the width of the pipe.

The situation is mirror-reversed for a moving pipe being pulled through stationary water. The water closest to the pipe's walls is dragged along with the pipe, the water right next to this ring moves a little slower, and so on. But in either case, for either the water moving through the pipe or the pipe moving through the water, the water right next to the pipe is stationary, *relative to the pipe*. As long as the flow is laminar, then right next to a moving object, there will be a thin layer of air (the arguments with a fluid such as water apply for a fluid such as air as well) that is not moving relative to the object. Just as in the example of the water faucet, this laminar no-slip zone is more robust the slower the motion through the fluid. At too high a speed the transfer of energy across concentric rings becomes disordered, and turbulence sets in. An object moving at a given speed must expend more energy generating turbulent flow than if the flow is laminar.

This is one reason why golf balls have dimples. The bumps on the golf ball decrease the cross-section of the turbulent wake behind the ball moving at higher speeds. In a crude sense, the dimples reduce the drag on the ball, as less energy is lost in the smaller turbulent wake. This effect was discovered accidentally. In the mid-1800s, golf balls were smooth, solid spheres of gutta-percha gum. It was empirically noted by golfers that old, scuffed-up balls with scratches and dings went farther for a given swing than brand-new, smooth balls. Experimental study and a theoretical understanding of fluid mechanics led to the optimal design of dimpled golf balls.

What's good for a golf ball is good for the Flash. As the Scarlet Speedster runs, the layer of air right next to him remains stationary relative to his body, so he has a non-moving pocket of air that he carries around with him at all times. Even in a layer only a few centimeters thick, there are nearly a trillion trillion O_2 molecules.

This "reservoir" of air must be continually refreshed with new air from outside the boundary-layer in order for the Scarlet Speedster to run for more than a few seconds at a time.

In *Flash* comics the no-slip zone "aura" that surrounds the Scarlet Speedster not only enables him to breathe as he runs, but also frees him from other untidy consequences of fluid drag. For example, if a meteor burns up in the atmosphere when it turns into a meteorite, due to the extreme frictional forces it experiences as it pushes the air out of its way entering the atmosphere at high velocity,* then why doesn't the Flash burn up when he runs at high speeds?

Flash comics # 167 provided an answer to this question, but it was a solution that few comic fans found satisfying. According to this story, the "protective aura" the Flash gained along with his super-speed powers was provided by a "tenth-dimensional elf novice-order" named Mopee. In this story, using his magical abilities, Mopee (who bore more than a passing resemblance to Woody Allen) removed the Flash's aura but not his super-speed. Consequently the Flash could still run at great speeds, but could not avoid burning up due to the tremendous air resistance he encountered as he ran.

That the Flash had an imp who bedeviled him was not as surprising as the fact that the Silver Age Flash went sixty-two issues in his own comic before encountering him. In the fifties and sixties, it seemed like nearly every superhero published by DC Comics had his or her own extradimensional pest. The first such character was Mr. Mxyzptlk, a fifth-dimensional being against whose magical powers Superman was powerless. Mxyzptlk could only be forcibly returned to the fifth dimension if he was tricked into saying his name backward, after which he was unable to return to our three-dimensional world for at least three months (presumably so that readers would not grow overly tired of him and his appearances would be noteworthy). Not to be outdone by the Man of Steel, Batman had his own magical imp, named Bat-Mite, whose attempts to honor his idol, the Caped Crusader, frequently backfired and

---

* Not all meteors burn upon entering the atmosphere, however. In order to account for the large quantity of Kryptonite that managed to reach Earth intact, it was argued in *Superman* # 130 that remnants of his planet's destruction possessed an invulnerability to air friction.

created mayhem and difficulties for Batman and Robin. J'onn J'onnz, the Martian Manhunter, had an alien sidekick named Zook while Aquaman had an imp named Quisp. Of the seven founding members of the Justice League of America, only Green Lantern and Wonder Woman have never had a supernatural or extradimensional spirit to call their own.

It wasn't the fact that the Flash finally acquired his own such imp that upset comic-book fans, but rather that Mopee claimed to have used his magical powers to give Barry Allen his super-speed powers. The science-fiction aspects that writers John Broome, Gardner Fox, Robert Kanigher, and editor Julie Schwartz introduced into the Silver Age with the creation of the Barry Allen Flash seemed undermined by the claim that the Flash's powers were in fact "magically" derived. Mopee never had a return engagement in *Flash* comics and, as far as most fans of the Silver Age are concerned, *Flash* # 167 never happened.

12

THE CASE OF THE MISSING WORK—

THE THREE LAWS OF THERMODYNAMICS

UNLIKE HENRY PYM, who found that he could re-
verse the polarity of "Pym particles" and enlarge himself into
Giant-Man, and the actress Rita Farr (Elasti-Girl) of the Doom
Patrol who, after breathing mysterious vapors from a previously
undiscovered volcanic vent during an African movie shoot, was
able to either grow to the size of a five-story building or shrink to
the size of an insect, the DC Comics hero the Atom could only
change his size in one direction: down.

The Atom is one of my favorite comic-book superheroes, for in
his secret identity he is Ray Palmer, physics professor. *Showcase*
34 introduced us to Professor Palmer as he was fruitlessly trying to
develop a shrinking ray, motivated by the economic benefits of such
a device. As he records in his audio journal after Experiment # 145
fails (well, actually, Experiment # 145 succeeded in reducing the size
of a kitchen chair to only a few inches, but the chair subsequently
exploded, just as every other object that had been shrunken), "compression of matter [. . .] would enable farmers to grow a thousand
times more than they do, on the same land! A single freight car
could transport the goods of a hundred freight trains." Of course, the
exploding part would make maintaining inventory stock difficult.

Palmer obtained the solution to his shrinking ray problems one
night when, on a nighttime drive, he spied a chunk of a white dwarf
star falling near him. This extraterrestrial material would turn out
to be the crucial missing ingredient that enabled Palmer to safely
miniaturize objects without subsequent explosion. Presumably
because this supernova remnant has the word "dwarf" in its name,

it possesses reducing properties previously unsuspected by physicists. We'll have more to say about the Atom's origin and the white dwarf meteor in a later chapter. Suffice it to say that the mechanism for how Ray Palmer was able to reduce his height to six inches (his typical crime-busting size) and much further, all the way down to smaller than an electron, makes about as much physical sense as inhaling strange underground vapors or Pym particles.

What is significant about the Atom was that unlike Ant-Man or Elasti-Girl, he was *not* restricted to a constant-density reduction. That is, the Atom could independently control both his size and his mass. Apparently, white dwarf matter's great density provides two "miracle-exceptions" for the price of one.

At normal height Ray Palmer was six feet tall and weighed approximately 180 pounds, equivalent to a mass of 82 kilograms. When reduced to a height of six inches, he is twelve times smaller than normal. His width and breadth will decrease also by a factor of twelve—that is, if he wants to avoid looking like his reflection in a fun house mirror. His volume therefore decreases by a multiple of $12 \times 12 \times 12$, or a factor of 1,728. If he were to shrink at a constant density, then his mass would also have to be reduced by this same factor of 1,728, leading to the Atom having a mass of only 47 grams, or one and two-thirds ounces. This is pretty light, and it is hard to see how such an ephemeral crime-fighter could tangle with the likes of Chronos the Time Thief or Doctor Light. Fortunately for those of us on the side of Good, the Atom could maintain his diminutive stature and increase his body weight up to his full-grown 180 pounds with just a click of his "size and weight controls" that he kept in his costume's belt buckle (later he added controls to his gloves as well). As shown in fig. 17, at his lighter weight he could use a pink rubber eraser as a trampoline, propelling himself at the face of a crook, and at the moment of impact increase his weight to 180 pounds. The resulting blow landing on the crook's chin would have the same force as when the Tiny Titan was his full height.*

Another trick the Atom would employ was, while only a few inches tall, to grab hold of the necktie that a criminal was wearing,

---

* Since Ant-Man shrinks at constant density, the force of his punch diminishes at the same rate as the cross-sectional area of his biceps. His punches pack an impressive pressure only because the surface area of his fist decreases along with the force supplied by his muscles.

Fig. 17. *Image from the cover of* Atom *# 4 where the Mighty Mite demonstrates his mastery of both his size and mass. Decreasing his density so that he is lightweight enough to use a pink rubber eraser as a trampoline brings him to within striking distance of a crook's chin. At this point he increases his mass dramatically, so that his punch kayos the "innocent thief."*

and then increase his weight to 180 pounds. The crook's head would be violently jerked downward until crashing on a tabletop or some other hard surface, knocking him out cold. This is undoubtedly one reason why criminals abandoned their formal attire later in the 1960s and were in the fashion avant-garde in adopting a more "casual Friday" approach to their work clothes when breaking the law.

Being able to control his mass (but only when his size was reduced from his normal height), the Atom could ride air currents generated by the wind or thermal gradients to get from place to place. (He would also make use of the fact that he could reduce himself to the size of an electron and telephone himself to distant locations—we will discuss the physics underlying this trick in

Fig. 18. *Panel from* Atom *# 2, wherein the Tiny Titan utilizes his ability to re-duce both his size and his weight to ride the thermal air currents from a burn-ing building.*

chapter 24.) The Atom would often be shown, as in fig. 18, gracefully gliding along air currents. In this panel, from *Atom # 2*, the Atom needs to get to the top of a barn that is on fire. He shrinks down to less than an inch and adjusts his density so that he is "lighter than a feather"—and then is pushed aloft by the "hot air currents to the roof." In reality, he should suffer a much more disorienting and punishing trip, and would, via direct, firsthand experience, verify the statistical mechanical underpinnings of the branch of physics called thermodynamics.

THE FIRST LAW–YOU CAN NEVER WIN

Just as Isaac Newton articulated three laws of motion, there are three laws of thermodynamics—the study of the flow of heat. The field of thermodynamics was empirically developed by scientists in the nineteenth century, well before they properly understood the atomic nature of matter. As such they came slowly to these laws, struggling with such problems as the connection between useful work and heat, a quantitative concept of temperature, the nature of phase transitions, and the intrinsic inefficiency of any mechanical process. Our consideration of this topic hopefully will be less painful, as we begin by elucidating the First Law of Thermodynamics with the simple question of why there is a net upward force on the Atom (fig. 18) as he rides the air currents created by the burning barn.

Back in the 1800s, scientists believed that matter contained a separate fluid termed "caloric" (previously named "phlogiston"). When mechanically deformed, objects would release their caloric fluid, and thus would feel warm to the touch or would expand when warmed, as the caloric fluid was supposed to be "self-repelling." It may sound silly today, but back before it was understood that all matter is composed of atoms, the caloric model seemed a reasonable way to account for these and other observations. The mystery of the true nature of heat was solved by Joule, Benjamin Thompson, and others, who demonstrated that mechanical work could be directly converted to heat, without the release of any special substance. The term "heat" describes any exchange of energy without the system performing Work (defined as force multiplied by distance). That any change in the energy of an object can only come about by either a transfer of heat or the performance of Work is the essence of the First Law of Thermodynamics.

Knowing that all matter is composed of atoms, we now recognize that when an object is "hot," the kinetic energy of the constituent atoms is large, while when an object is "cold," the kinetic energy of the atoms is lower. The "temperature" of an object is therefore just a useful bookkeeping device to keep track of the average energy per atom of the object. We say something has a high temperature when the atoms contain a larger amount of kinetic energy compared to another object with a lower kinetic energy per atom, which we call "cold."

The rubbing of two objects against each other, which, in our earlier discussion of friction we noted can be viewed at the atomic level as the scraping of one atomic mountain range across the other, results in the transfer of kinetic energy (the macroscopic motion of the top object being dragged) into the shaking of the constituent atoms in each surface. This transfer of kinetic energy per atom is referred to as "heat," and the temperature of the respective objects is raised, without requiring a caloric fluid be present.

From chapter 11 we know that energy comes in only two flavors: Potential or Kinetic. As the mass of any given air molecule is very small, the variation of the gravitational potential energy (given by the weight of the molecule multiplied by its distance above the floor) across the height of the room is so tiny that we make only a very small error by ignoring it. Thus the main component of the average energy of the air molecules in the room is kinetic energy. Since the following discussion does not depend on the exact chemical composition of the atmosphere, we'll refer to generic "air" molecules. The hot air underneath the Atom in fig. 18 has a greater kinetic energy than the colder air above him. Of course, this last statement is unscientific, for there is no such thing as "hot air" or "cold air," just air that is hot compared to some other object or cold compared to another reference, which for the sake of argument we'll take to be the Atom's body temperature.

In addition to pushing him above the burning barn, the hot air beneath the Atom in fig. 18 warms him up. Usually when a speeding car collides with a stationary or slowly moving vehicle, the faster car slows down and the slower car speeds up. Similarly air molecules that have a higher average energy than those in the Atom's body will, upon colliding with him, transfer on average some of their kinetic energy to the atoms in his costume, causing them to shake more violently than before. Since energy is always conserved in any process, the average energy (that is, the temperature) of the air will decrease following the collisions, while the temperature of the Mighty Mite will increase through this transfer of kinetic energy. The atomic nature of matter makes clear why, when two objects are in thermal contact, the net flow of heat is always from the higher-temperature object to the lower-temperature object, and never in the other direction. Similarly, when exposed to cold air, the atoms in the Tiny Titan's costume will be moving back and forth much faster than the air molecules that they collide with, so after the collision the air will be moving faster at the expense of the atoms in his costume, cooling the Atom down.

When the Atom is lightweight enough to float on air currents (as illustrated in fig. 18), and if all of the air surrounding him is at the same temperature, he is being bombarded in all directions with roughly equal force. In this case no matter how small his mass, eventually gravity will pull him down to the ground. The thermal draft that keeps the Atom aloft arises from the hotter air molecules below the Atom that are moving faster than the cooler molecules above him. There are therefore more collisions per second beneath the Atom that push him upward than collisions above him that direct him toward the ground. Furthermore, the force needed to reverse the direction of a faster-moving molecule is larger than if the molecule is moving slowly. Forces come in pairs, so the force the Atom exerts on the air molecules, changing their direction, also pushes back on the Atom. There will therefore be a net, unbalanced force on the Atom due to there being more collisions underneath him pushing him upward than above him driving him down. Of course, this force will not be smooth and uniform, but will in fact be discontinuous and noisy, and occasional statistical fluctuations will lead to him being pushed down rather than up, but over time the average force from the thermal gradient will be to push him away from the hot source and toward the cooler region.

The Atom in fig. 18 therefore gains both kinetic and potential energy. He warms up slightly (that is, the kinetic energy of the atoms in his body increase) due to the flow of heat from the hot-air current. He will also be raised to a greater altitude, increasing his gravitational potential energy, thanks to the Work done on him by the net force exerted by the hotter air molecules. The First Law of Thermodynamics states that the net change in the Atom's total energy is the sum of the heat flow and the Work done on him.

Another example: When a hot gas pushes against a piston in an automobile cylinder, lifting the piston and thereby, through a series of cams and shafts, causing the wheels to rotate, the energy of the hot gas that goes into moving the piston is termed "Work." The Industrial Revolution came about when scientists and engineers realized that during the transfer of energy through the flow of heat from hot to cold, productive work, meaning a force applied across a given distance, could be extracted. Prior to this, the development of simple machines such as levers and pulleys to amplify forces required the energy stored within humans, draft animals, wind, or waterfalls. The Work supplied by humans or animals is converted from the stored chemical energy in the foodstuffs ingested. The potential energy in the food has to accomplish many other tasks, from maintaining body temperature, to continuing metabolic functions and so on. Consequently the amount of energy available for pushing down on a lever is only a small fraction of that stored within the food. In contrast, releasing the stored potential energy by burning coal or oil enables the more direct conversion of the resulting kinetic energy into work. While not 100 percent efficient, it is far more advantageous than using living beings.

The First Law of Thermodynamics tells us that in a best-case scenario, with all losses and external noise removed, the total amount of work we can get out of any device is exactly equal to the heat flow (change in kinetic energy) that drives the machine. By the principle of conservation of energy, it is impossible to extract more work than is available from the heat flow from a hot to cold source. The heading of this section implies that the universe ensures that you can never win (that is, get more out than you put in).

Well, if winning is ruled out, why is breaking even so difficult? Couldn't we ever make a perfect machine that, once started, would continue indefinitely without further fueling? We'll see in a moment that the random motion of the gas atoms during the transfer of heat places strict limits on the amount of usable work we can obtain from any machine, no matter how cleverly designed.

THE SECOND LAW—YOU CAN'T EVEN BREAK EVEN

There is nothing in the principle of conservation of energy, which underlies the First Law of Thermodynamics, that prevents or forbids the construction of a 100 percent efficient machine, where the work created by the device exactly equals the heat energy put into it. In fact, if all we had to go on was the First Law of Thermodynamics, we would reasonably expect that machines *must* be 100 percent efficient, for we know that energy can never be gained or lost, but only transformed from one form to another. In order to understand what limits the conversion of heat into work, we must introduce a new concept, complementary to, but as important as, energy. This concept, called "entropy," is intimately connected to heat flow, and will give the Atom a very bumpy ride, even when he isn't floating along thermal drafts.

Whenever there is an explosion on the Justice League of America's satellite headquarters orbiting in outer space, or on the quinjet spaceship used by the Avengers, there is a violent outpouring of air into the low-pressure surroundings. Why? What compels the air to rush out of the opening in the JLA satellite? A common metaphor invoked to explain why air races into any region that is at a lower pressure is "nature abhors a vacuum." And yet there would be no violation of Newton's laws of motion if all the air were to remain inside the JLA's satellite, even with the door left open, though admittedly such a situation is extremely unlikely. It is the random motion of the air associated with the air molecules' kinetic energy that underlies the explosive decompression on the satellite.

Imagine that the room next door to the one you are sitting in has had all of its air removed. As long as the door connecting the two rooms is kept hermetically sealed, you would never know that a perfect vacuum was waiting in the next room. The air molecules in your room are at a certain temperature and pressure, and are buzzing merrily along. This peaceful, stable scenario changes when the door separating your room from the vacuum room is opened (with the door swinging into the evacuated room).

Instead of asking why the air would rush out of your room into the vacuum room once the door was opened, the better question is, why wouldn't it? Those air molecules that were moving toward the door, and would have bounced off of it had it remained closed, will now continue moving in a straight line into the vacuum room (per Newton's first law). However, only a small fraction of the air molecules in your room would have been heading toward the door right before it was opened. Some of the air molecules would be moving away from the door, where they would collide with other air molecules moving in different directions. It is conceivable, though implausible, that aside from those air molecules initially heading toward the now-open doorway, all of the remaining air molecules would continue to collide with one another, and no more will pass into the second room. This is as likely as, with the door remaining closed, all of the air molecules, through random collisions, managing to always avoid the region right next to the door. You would not have to worry about suffocating if you sat next to the door, because a fairly constant fraction of the air is always heading toward you at any given moment. The air molecules occupy all regions available to them—there is no reason for them not to. Any particular air molecule may spend most of its time in one particular corner of the room, but on average every volume is just as likely to have air in it as any other, just as in a well-shuffled deck of cards, any one of the fifty-two cards is as likely to be turned up at the top of the deck as any other.

Air molecules have no free will, and if they are going to collide and head toward the door when it is closed, they will do so when the door is open. The only difference, and it is a big difference, is that once the air molecules move into the vacuum room, there are initially no other air molecules for them to collide with. There are many, many more ways for the atoms in the first room to move into the vacuum room, than for them to bounce off one another and never pass into the second room. "Entropy" is the term used to describe the number of different ways a given system can arrange itself. A brand-new deck of cards with all the cards arranged in numerical order by suit has low entropy, while the entropy is at a maximum once the deck has been thoroughly shuffled. It is hard to be dealt four aces in a row from a well-shuffled deck, just as it is harder to know where a particular air molecule is if it can possibly be in two rooms rather than just one.

It is not impossible—it would violate no law of physics—for all of the molecules in the room you are sitting in right now to, as a result of random collisions, move over to the side of the room in which you are not sitting. But don't worry; it is more likely that an honest shuffling of a mixed deck of cards will result in the cards being arranged numerically by suit as when newly purchased. The odds of the air molecules moving to one side of the room are so low that you would have to wait longer than the age of the universe before you are likely to see this occur. When physicists say that systems tend toward maximum entropy, all they are saying is that the most probable situations will be the ones observed. When you pull your socks out of the dryer after they have been tumbling for some time, it is possible that you will extract them two at a time in perfectly matched pairs, but don't count on it. There is only one way that could happen, but many more ways in which the socks can be mismatched, so the most likely result will be that you will have to sort the socks later. The dryer randomizes the socks, so every sock has an opportunity of pairing with any other sock. Issues concerning entropy only apply to such fully randomized situations, and there are obviously many examples of externally imposing order on a system (such as manually sorting the socks).

When the door between rooms is opened and the atoms move into the second, initially vacant room, we say that the entropy of the air molecules increases, but what this basically means is that if something can happen, it will happen. There are many more ways for the air molecules in the room to be spread out uniformly, each sharing a nearly equal portion of the total kinetic energy of the room, than for all the air to collect in one corner, or for one molecule to have all the kinetic energy of the atmosphere and the rest of the molecules to have none. Those things that are observed most often will be those most likely to occur. The mutant Scarlet Witch, originally a villain in Magneto's Brotherhood of Evil Mutants in Marvel Comics' *X-Men* and later rehabilitated as a hero in the *Avengers*, had a "hex power," whereby she could gesture at an object and something untoward would occur. It was suggested in *West Coast Avengers* # 42 that her power consisted of the ability to alter probabilities, so that a highly unlikely event would actually have a near certain probability of happening. Based upon the previous discussion, we would further characterize her mutant talent as the power to alter the entropy of a system, bringing about rare configurations (such as all the air moving over to one side of a room) much sooner than would reasonably be expected.*

In order to get useful work out of the compressed gas in a car cylinder or the steam in a boiler, it typically must expand from a confined region to a more spacious configuration. Think about the air molecules in the room, separated from the vacuum room. It would be nice if they could be arranged to push the door open on their own, without your having to do so manually. If we unlock the door between the two rooms, then there are many air molecules striking the door on one side, and none counteracting on the other, so that there will indeed be a net force on the door that could push it open. This is the same unbalanced force that pushes the Justice League members out of their satellite headquarters when there is a hull breach. This unbalanced force arises quickly—the time between collisions for air molecules at room temperature is less than a nanosecond (one billionth of a second).

Once the air has moved into the vacuum room, it has a higher degree of disorder, that is, its entropy has increased. Closing the door, the air will not again push the door open, unless I scoop all the air out of the second room, repeating the original configuration (one room at atmospheric pressure, the other room under vacuum). This takes effort, and when I count up all the energy expended in returning the system to its original state, I wind up

---

* Years later Wanda's mutant power was associated with "chaos magic," but this is just fantasy. No less an authority than Dr. Strange (the Sorcerer Supreme of the Marvel universe) has stated (*Avengers* # 503) that there is no such thing as "chaos magic."

using more energy than I gained when the door was pushed open. No matter how cleverly I arrange things, I can never convert all the energy of a system into useful work—there will always be some part that goes into increasing the entropy, which is not useful.

When the mixture of gasoline and oxygen is ignited in your automobile engine, it undergoes a chemical reaction, releasing heat (that is, the reactant products are moving faster than they had been before the explosive reaction). Only those faster-moving molecules that are heading in the right direction will displace the piston in the engine, leading to rotation of the tires. Those molecules that are heading away from the piston are wasted, from the point of view of getting something useful out of the chemical reaction. Not only can you not get more energy out of a process than you put in, but the entropy constraint means that you will always get out *less* useful work than you used to set the system up. This is the heart of the Second Law of Thermodynamics. No process can be 100 percent efficient, and in point of fact most motors and engines rarely convert more than a third of their energy into useful work. The Second Law of Thermodynamics is a harsh mistress, but there doesn't seem to be any way around it. Or is there?

Could I use the talents of the Atom to try to beat the Second Law of Thermodynamics? The air molecules in your room are characterized by a certain temperature, which measures the average energy of the air. The key word is "average"—not every single air molecule in the room has exactly the same kinetic energy. Some are moving a little faster than average while some poke along a little slower. The steam coming off a fresh cup of coffee is a reflection of the fact that not every molecule in a collection has exactly the same energy. Those water molecules in the coffee that are energetic enough to escape from the liquid state (more on such phase transitions in chapter 14) form the clouds hovering over the coffee. The hotter the coffee is initially, the greater the steam forming over the liquid surface, as there are more water molecules at the high end of the kinetic energy distribution to break out of the liquid state. When you blow on your coffee to cool it off, you do not reduce the temperature of the coffee because your breath is a frosty 98.6 degrees Fahrenheit. Rather what you do is disturb the steam, pushing the most energetic water molecules away so that they are unable to be deflected back into the coffee. Once they are permanently removed from the coffee/steam system, the average energy (that is, temperature) of the remaining coffee decreases. This physical process is called "evaporation cooling" and is the process underlying the operation of refrigerators and is the reason why sweating is more effective in cooling you down if there is a strong breeze to carry the perspiration away.

The idea of using the Atom to get around the Second Law of Thermodynamics is a variation on the concept of evaporation cooling. We'll start by having the Atom shrink down so that he is only several times bigger than an air molecule. He'll have a box with him, with a small, hinged door. In this example the Atom assumes the identity of "Maxwell's Demon," proposed by James Clerk Maxwell to test the Second Law of Thermodynamics. All the air molecules in the room are at the same temperature, which means that they cannot be used to generate a heat flow to power a machine. But now the Atom makes use of the fact that the temperature is an average measurement, and starts sorting out the air molecules, based upon their kinetic energy. Those air molecules coming toward him that are moving faster than average, he collects by opening the door of his box and trapping them inside (it's a thermally insulated box, so those molecules retain their kinetic energy once secured inside). Those that are moving slower than average, he ignores. Before long he has acquired a large number of air molecules that have a kinetic energy larger than the corresponding initial average value. Furthermore, by removing these faster-moving molecules, the average energy of the remaining air molecules decreases, just as when you blow on your steaming coffee. The Atom can now take these hotter-than-average molecules and, bringing them into thermal contact with some colder molecules, allow the net heat flow between them to power an engine, thereby getting useful work out of air that was initially at one average temperature.

Or he would, if we didn't have to worry about the Atom himself. He expends energy in opening and closing his box to sort and trap the energetic molecules. This energy must be included in any balancing of the total energy put into and extracted from a process. To neglect his contribution would be equivalent to saying that you are able to drive your car for just pennies a day, if you ignore the cost of the gasoline. When the heat and work contributions of the Atom's sorting of the air molecules are carefully accounted for, we find that by collecting the faster molecules, the Atom himself contributes energy to the remaining atmosphere, increasing its average kinetic energy, so that in the end there has been no net temperature differential. If you blow on your coffee and remove the steam but replace it with other molecules that are just as hot, you have not cooled your drink.

No matter how hard you try (and believe me, many have tried), there is only one way, discussed below, around the no-win scenario presented by the Second Law of Thermodynamics. Unfortunately, even that option is not available to us.

THE THIRD LAW—YOU CAN NEVER GET OUT OF THE GAME

If entropy considerations limit the amount of useful work we can extract from any process, whether it's a V-8 engine, a gas turbine, or the chemical reactions in your cells' mitochondria, then couldn't we just get around this problem by dealing with systems with no entropy? After all, it is conceivable, no matter how difficult it may be in practice, to have a system where all the atoms are in a precise, uniform configuration, so that there is no uncertainty regarding the location of any single element within it. Why can't I arrange my two systems that generate the heat flow that powers my engine to have no entropy, so that I don't have to worry about the Second Law?

The reason why this won't work is that the entropy of a substance and its internal energy (which could be available for heat transfer) are related, such that we can't change one without affecting the other. The entropy of the air molecules in the room is a measure of their random motion. If I lower the air's kinetic energy, eventually the gas condenses into a liquid. The entropy of the liquid is lower than that of the same molecules in their vapor state, because there is less uncertainty as to where any given molecule might be (they're in the puddle on the floor, as opposed to spread out throughout the room). But there are still chaotic fluctuations in position and velocity of the molecules in the liquid state. Lower the temperature of the liquid further, and eventually the average kinetic energy of the molecules is insufficient to overcome the attractive bonding between molecules, and the material freezes into a solid. The chemical bonds between the molecules have preferred orientations, so the natural configuration of the solid will be a particular crystalline arrangement, with all of the atoms or molecules lined up in a certain way. At very low temperatures, all of the atoms will be in their ideal crystalline spots, and we will know the location of any given atom.

The entropy of any crystalline solid would therefore be zero, except for the atoms' vibrations about their crystalline positions. The solid will still have some temperature, no matter how low, so the atoms in the crystal will be shaking back and forth. We will have truly no uncertainty, and the entropy will be exactly zero, only when *all* vibrations of *every* atom in the solid cease. The fact that the entropy is zero only when the temperature is also zero is termed the Third Law of Thermodynamics. In the zero-temperature state none of the atoms have any kinetic energy at all. At this point we say that the solid has a temperature of Absolute zero degrees. We use the prefix Absolute, because no matter what type of thermometer you use, it will read zero average kinetic energy at this point. Note that not even outer space is this cold. There is background light and stray cosmic rays even in the vacuum of space, and they all carry energy. In fact, the radio-wave background radiation that is a remnant of the Big Bang origin of the universe has an energy characterized by an average temperature of 3 degrees above Absolute zero. So, even outer space has a temperature, and hence an entropy. The only way to beat the Second Law of Thermodynamics is to use systems with zero entropy, but this can only be realized at Absolute zero. But if everything is at zero degrees, how would there be any heat flow to power our engine? The three laws of thermodynamics almost conspire to prevent us from constructing perfectly efficient machines, and just like supervillains in a comic book, we must reconcile ourselves to inevitable loss.

Our discussion of entropy has relied so heavily on the fluctuations of the constituent atoms that it is striking that the Second Law was formulated long before most scientists were convinced that matter was indeed composed of atoms. From the mid-1800s onward, some scientists had been taking the atomic theory of matter more and more seriously, while others remained unconvinced of the reality of atoms. These critics felt that while the suggestion that matter was composed of atoms was a useful idea that simplified many of the calculations about the properties of fluids and gases, it was nonetheless meaningless to ascribe physical reality to entities that were too small to ever be seen.* Many of the elder statesmen of physics at the time, notably Ernst Mach (for whom the speed of sound in air, the Mach number, is named) held this view.

Nevertheless, the atomic hypothesis eventually won out, via the same strategy by which all revolutionary ideas succeed. As Max Planck, himself a young Turk of the quantum revolution (about whom we will have much more to say in section three), once remarked, "A new scientic truth does not triumph by convincing its opponents and making them see the light, but rather because its opponents eventually die and a new generation grows up that is familiar with it."

A key development that convinced these younger scientists that atoms were real, regardless of what the older establishment claimed, was in accounting for the jitter that small objects underwent due to their random bombardment from still smaller atoms and molecules striking them from all sides. This phenomenon is termed "Brownian motion" after Robert Brown, the botanist who observed the excursions of a pollen grain in a drop of water using a new scientific instrument, the microscope. While Brownian motion had been known since 1828, it was not until 1905 that a satisfactory theoretical description was provided in the Ph.D. thesis dissertation of another young upstart, Albert Einstein. Einstein was able to quantitatively calculate the excursions of a pollen particle due to the collisions with the water fluid in which it was suspended, and also relate the magnitude of the fluctuations to the temperature of the surrounding medium. The close agreement between Einstein's calculations and experimental observations convinced many physicists that the atomic hypothesis was indeed correct. While his thesis research was not as revolutionary as his subsequent investigations of relativity (published the same year), Einstein would have been well known to physicists even if his

---

* It's a good thing the Atom never had to hear such hurtful comments.

only contribution to science were his elucidation of the statistical nature underlying Brownian motion.

As the Atom shrinks down so that he is roughly the size of a pollen grain—say, a hundredth to a tenth of a millimeter, less than the diameter of a human hair—he will begin to experience the back-and-forth motion that Brown first observed. This size is critical: When he is larger, the average bombardment is negligible; when much smaller, he fits between the atoms in the air, so that as long as he can avoid being struck, he will be fine. At any given instant there may be more molecules striking the Atom from below so that he will suddenly be pushed upward, while the following moment may see a downward thrust, perhaps not as severe as the last upward swing, or possibly even more extreme. On average, the Atom will not experience a net displacement from this back and forth, but he will need some Dramamine before too long.

One doesn't have to be as small as the Atom to directly experience Brownian motion. The random collisions of the air on our eardrums produce deflections that are just at the limit of our hearing. Sit in a soundproof room for about thirty minutes and your hearing improves (just as your eyes' sensitivity to stray light increases when you've acclimated to a darkened room) until you'll be able to detect the deflection of your eardrums by the motion of the atoms. It is possible in a very quiet room to hear the background noise emanating from the entropy of the air, in essence to *hear* the temperature of the room. Super-hearing—it's not just for Kryptonians anymore!

13

MUTANT METEOROLOGY—

CONDUCTION AND CONVECTION

STAN LEE, head writer and editor of nearly all Marvel comics in the Silver Age, was fond of radiation as a source of his heroes' superpowers. The bite of a radioactive spider gave Peter Parker the proportionate strength and abilities of a spider; the Fantastic Four were bombarded by cosmic rays (high-energy protons from as near as the sun and as far as other galaxies); Bruce Banner turned into the Hulk when he was belted by gamma rays (more radiation); and Matt Murdock was struck in the eyes with a radioactive isotope that literally fell off a truck, blinding him but endowing him with a "radar sense" and enhancing his other senses so that he could fight crime as Daredevil. After all of this radioactivity, Lee was tired of trying to come up with such origins for bizarre superpowers. Hence in 1963 when he co-created with Jack Kirby a new team of superpowered teens, the X-Men, he essentially threw in the towel, claiming that they were mutants, and hence simply born with their strange abilities and attributes.

One of the original X-Men, first appearing in *X-Men* # 1,* was Bobby Drake, code-named Iceman. Drake's mutant power was the ability to lower the temperature of his body and his immediate

---

* Later to be known as *The Uncanny X-Men.*

surroundings to less than 32 degrees Fahrenheit. His body would thereby acquire a protective coating of frozen water. As explained in *X-Men* # 47, Bobby does not generate the ice that covers his body or projects from his hands when he employs his mutant power. Instead, by lowering the temperature around him, he condenses the water vapor that is always present in the air. By now I don't need to belabor the point that any heat—that is, any kinetic energy— that Bobby is able to subtract from his surroundings must be compensated for by an amount of heat added somewhere else and, given the Second Law of Thermodynamics, the heat added is most likely greater than the heat removed. Refrigerators remove heat from an enclosed space, but this heat must then be deposited elsewhere. In addition, the motors in the compressors of the refrigerator require energy, and some of this electrical energy is not converted into useful work but takes the form of "waste heat." The waste heat of a refrigerator is ejected from its rear; usually placed against a wall. (If you ever want to heat your kitchen, just leave the refrigerator door wide open. As the appliance struggles to lower the temperature of the room, it will deposit more heat into the kitchen than it is able to remove.) Where Iceman deposits the excess heat generated when he lowers the temperature of his surroundings remains a mystery.

Iceman's body covering initially took the form of fluffy snow, and when he had gained further control over his power in *X-Men* # 8, he acquired a crystalline, icy appearance. The difference between snow and ice is in the arrangement of the water molecules that result when the water freezes into the solid state. Snowflakes are constructed from aggregations of water within clouds. When water molecules condense from the vapor phase, they release energy, warming the surrounding air. The lower density warmed air keeps the cloud aloft, as in a hot air balloon. When too many water molecules come together, frequently coalescing around a dust grain in the cloud, they form a droplet. When the temperature in the cloud is above 32 degrees Fahrenheit, the droplet can fall from the cloud in the form of rain, converting its potential energy into kinetic energy. A frozen rain droplet is called sleet. The construction of a snowflake is a more delicate affair. A snowflake is created when water vapor slowly freezes around a dust particle. The water molecules pack into a hexagonal lattice, owing to their chemical shape. Metal atoms stack in a solid like cannonballs or oranges in a grocery display, where this close packing structure is determined by the nature of the chemical force holding the metal atoms together. Water molecules have a more V-shaped geometry, with an oxygen atom at the vertex bonded with two hydrogen atoms, protruding like a rabbit-ear antenna. The shape of the water molecule determines the geometry of its packing, which turns out to be a hexagon.

Chemical stacking explains the sixfold symmetry of snowflakes, but how do they form their characteristic lacy structure? In the snowflake-generating clouds that have a relatively low humidity, the water molecules must diffuse to the growing flake before they can be incorporated into its structure. The water molecules in the cloud are not being pushed in any given direction, but are undergoing Brownian motion as they fluctuate one way and another. This type of "random walk" is a very slow way to get anywhere, since you are just as likely to take a step away from your goal as toward it. When you hold your hand over a hot object, the warmth you feel is carried by air molecules randomly diffusing away from the high-temperature region. This method of conveying energy from one location to another is termed "conduction," and is fairly inefficient. In general, unless the object in question is glowing white hot, you typically have to place your hand very close before a significant energy transfer is detected.

Einstein's equation for how far a fluctuating atom moves as a function of time, derived in his 1905 paper on Brownian motion, indicates that it will take a hundred times longer for the water in a growing snowflake to diffuse a distance of one centimeter than it will take for it to travel one millimeter. Consequently, those regions of the growing snowflake that extend farther out from the body of the flake will accrete water faster, because they shorten the distance the molecules must random-walk to reach the flake. The six pointed regions at the corners of the hexagon will therefore grow first by adding water molecules, and as they extend farther, they continue to grow more rapidly than neighboring regions. The exact details of how the flake develops—how the dendritic branches grow secondary offshoots, the role that the energy deposited by the diffusing water molecules plays in locally melting and then refreezing the growing flake—will all depend sensitively on the humidity and temperature within the cloud. The final structure of the flake will also depend on the unique details of how the flake forms about a dust particle, such that no two snowflakes will ever be *exactly* alike, though it is possible to find flakes that are strikingly similar. But at its heart, the beautiful symmetry and order of a snowflake arises from the disordered fluctuations underlying Brownian motion.

* * *

As his mastery of his mutant ability progressed, Iceman was able to project "freeze rays" from his hands, icing up another person or object, or even creating a large mountain of ice beneath his feet. To get around while fighting such evil mutants as Magneto, the Blob, or the sinister robotic Sentinels, Bobby would frequently generate an "ice slide" underneath his feet on which he would glide, as illustrated in fig. 19. Bobby would in principle create a large ice mountain underneath himself and then generate a ramp on which he would slide to his intended destination. Well, this in itself would not violate any physics principles, assuming of course that one could indeed control local temperatures in this manner and that there was sufficient humidity in the air to create all this ice. What is troubling however is the apparent stability of Bobby's ice slides, no matter how extended they became. At some point Bobby would get out in front of the center of mass of his ice slide, at which point one would expect bad things to happen.

The "center of mass," also referred to as the center of gravity, is the spot at which an object behaves as if all of its mass, no matter how much or how unevenly distributed, were concentrated at this one location. A yardstick would have its center of mass located exactly at its center. You can balance the yardstick across your index finger, held so that it lies parallel to the ground, but only if your finger were underneath this midpoint. Placing your supporting finger too close to either end will cause the stick to rotate and fall. The center of mass depends on the distribution of matter in the object. A baseball bat, thicker and heavier at one end, will have its center of mass closer to the wider end than the narrower handle.

To see why Bobby's ice slides cannot extend too far without

Fig. 19. *Scene from* Amazing Spider-Man # 92, *where the mutant X-Man Ice-man battles Spider-Man due to a mistaken impression (such confusion that would lead two heroes to fight until they realized that they were on the same side occurred no more often in Marvel comics than once or twice—a month). It's no surprise that Spidey has never seen anything like Iceman's ice ramps, as they display questionable stability when the mutant hero ventures far beyond the ramps' center of gravity.*

crumbling, place a book on a table. The weight of the book is directed toward the ground and is balanced by a force from the table. The center of mass is in the middle of the face of the book, and as long as it is on top of the table, the book is stable. But now slide the book near the edge of the table. At first a small fraction of the book can hang over the edge of the table with no problem. The portion of the book that's over the edge creates a twisting force, a torque. This unsupported weight of the book tries to rotate the book, but there is more weight over the table, trying to rotate the book in the other direction, so the book remains stationary. But as you slide the book farther, so that its center of mass is no longer above the table but over the side, then the book will rotate and flip onto the floor. This is because the torque that is trying to rotate the book off the table is now greater than the counter-torque that is trying to keep the book on the table.

Similarly when Iceman's slides become too extended, he moves too far from the center of mass of the ice mountain he'd originally generated. The torque that results from him gliding along its edge becomes greater than the strength of the ice slide, and his slide should break off, just as the book will flip off the table when too much of its weight is over the edge. In order to satisfy basic mechanics, Bobby Drake should continually reinforce the underside of his

Fig. 20. *A scene from* X-Men *# 47, where Bobby Drake (Iceman) talks directly to the reader about the mechanisms underlying his mutant freezing powers and gives a knowing wink that he recognizes the physical implausibility of some (OK, all) of his uncanny feats.*

ice slides with ice pillars, in order to avoid getting too far in front of their center of mass. Sometimes the implausible mechanical stability of Iceman's constructs would be acknowledged, as indicated in fig. 20 from a backup feature "I, the Iceman" in *X-Men* # 47. If any "wiseguy physics majors" wonder what keeps Iceman's ladders and ramps aloft, the answer, which no one can refute, is "a hulkin' helping of imagination!"

* * *

The X-Men comic was not a major sales success in its first incarnation in the 1960s, and original stories ceased publication in 1970. Five years later a new management at Marvel Comics decided to try a revival of the X-Men, and the *All-New, All-Different X-Men* (written by Len Wein and drawn by Dave Cockrum), as their debut giant-size issue was titled, was a financial success from the very beginning. Despite the promise implied by the title, some of the X-Men characters featured in this issue were from the original team, but many of the new characters debuting, such as Ororo Munroe (Storm); Logan (Wolverine); Peter Rasputin (Colossus); and Kurt Wagner (Nightcrawler) would go on to become fan favorites. Not everyone was impressed, however. Stan Lee has been quoted as complaining about the implausibility of Storm's mutant ability to control the weather. He did not seem bothered by Colossus's ability to transform his skin into "organic steel" (I have absolutely no idea what this phrase means, or how this power could possibly work in the physical world) or Nightcrawler's teleportation power (well, if you give me enough "miracle exceptions" I suppose I could try to make this work), but "weather-powers" seemed too implausible for the man who brought the world the Incredible Hulk and the Silver Surfer (if he travels through outer space, what exactly is he surfing on?).* But Stan should not be too quick to throw stones at superhero houses. The same mutant power manifested by one of Stan's own creations, Iceman of the original X-Men—the ability to generate and control thermal gradients—also enables Storm to influence meteorological phenomena.

At its core, the weather is simply a matter of the atmosphere absorbing energy in the form of sunlight, and is in fact so simple that it is nearly impossible to accurately predict. When one thinks of weather, terms such as "wind," "rain," and "snow" come to mind (particularly if you live in Minnesota, like me). All are governed by spatial differences in temperature, driven by variations in the sunlight energy absorbed by the atmosphere.

Spatial variations of the atmospheric temperature are associated with changes in the atmospheric density (the number of air molecules in a given volume). When a volume of denser air is adjacent to a dilute region, there will be a net flow of air from the high- to low-density spaces until the density in each volume element is equal. This flow of air can be understood simply on the basis of the entropy argument discussed in the previous chapter. If there is a constant input of energy, keeping one region at a lower density than another, then this flow of air—or wind—will persist. The wind can in turn move cloud cover, changing the spatial pattern of sunlight absorption, which changes the air-

---

* Cosmic Rays?

flow trajectories that influence the cloud cover, and so on. Of course, the Earth's rotation determines the global direction of air circulation.

The ability to accurately predict the weather is therefore limited by the precision with which one knows the initial air speeds and temperatures as a function of space and time. Furthermore, changes in temperature lead to air flow that in turn changes the absorbed sunlight, leading to new air-flow patterns. There is a nonlinear feedback in place, such that any small uncertainty in our knowledge of the initial conditions quickly becomes magnified. In a linear system a small change in the input results in a corresponding small variation in the output, while for nonlinear systems, such as the weather, a small change could lead to a large variation in the output. This has become popularly known as "the butterfly effect," whereby the beating of a butterfly's wings in Cleveland can, several weeks hence, produce tornadolike conditions in Chile. Meteorologists can do an excellent job of predicting the weather in the short term, but anything beyond a few weeks is intrinsically unreliable, regardless of the quality of the measurement systems.

A physically plausible explanation for Storm's ability to control the weather is that she is able to alter atmospheric temperature variations in space and time at will. The wind that allows Storm to fly, as illustrated in fig. 21, is created by a temperature gradient beneath her. Storm presumably uses her mutant power to make the region of air under her hotter than that above her. The temperature of the air is a measure of its average kinetic energy, so air at a very low temperature is hardly moving at all—nearly all of its energy is gravitational potential energy, and it will fall toward the ground. The faster-moving "hot" air molecules will occupy the space left vacant by the falling "cold" air molecules, simply because if they are zipping around at great speeds, colliding with one another, there are many more ways they can scatter into unoccupied regions than if they collide and always manage to stay near the ground. The average kinetic energy of the hot air molecules is large; consequently, the gravitational potential energy is only a small addition to its total energy. Once the hot air molecules are near the cold upper region, and the cold air molecules are near the ground, the lower molecules will gain energy from collisions with the heated

Fig. 21. *Scene from X-Men # 145, where the mutant Storm employs her ability to generate controlled thermal gradients to warp wind patterns, carrying her aloft on convection cells.*

ground, and the hot air will lose energy following collisions with the cool air above it. There will once again be a situation where hot air is on the ground and cold air above it, and the cycle will continue.

This process is termed "convection" and such thermal convection rolls are an extremely efficient way to transfer energy from a hot source to a cold one through the thermal link of the air in the room. In fact, this is why double-paned windows provide good thermal insulation: By physically separating the glass inside the room and the cold pane facing the outdoors, the windows keep the interior glass closer to room temperature. The cold outer window is unable to set up a strong convective roll within the room. Of course, energy can still be transferred from air molecules colliding with the warm inner glass, and then depositing this energy to the colder outer glass window, but this conduction is much slower than convection.

The fraction of air that is composed of water vapor depends on the average kinetic energy (ambient temperature) and pressure of the atmospheric molecules. Colder air is denser and has less room to accommodate the water molecules. If Storm is indeed able to control the local temperature, then she can also vary the barometric pressure and humidity at will. It is not unreasonable that she would be able to cause localized rain- or snowstorms, or even generate lightning strikes, though her ability to control the exact position of the strike would be hampered by factors (such as the local charge buildup on the ground) outside of her control. All told, if Stan Lee was not bothered by his creation of Iceman, a mutant who could lower his own body temperature below 32 degrees Fahrenheit and also project localized regions of lower temperature in his immediate vicinity, then a mutant such as Storm who can control not her own temperature but that of the surrounding atmosphere should not be too great a stretch.

* * *

A final thought about a connection between genetic mutation and thermodynamics. According to Stan Lee, mutants, particularly those with dramatic superpowers, comprise an entirely new species, Homo Superior, and are distinct from most comic-book readers, Homo Sapiens. The process of speciation, by which new species develop, was put forth by Charles Darwin and independently by Alfred Wallace in the 1850s. In Darwin's original formulation of the theory of evolution, he proposed that speciation was a slow, gradual process and that several hundred million years were necessary to account for the current biological diversity. The only problem was that the physics of the time provided an estimate of the age of the Earth, and it was only roughly twenty million years old.

One of the foremost scientists of the nineteenth century, William Thomson (later honored as Lord Kelvin for his efforts in developing transatlantic telegraph cables) performed a thermal conductivity calculation that challenged Darwin's hypothesis. Thermal conductivity is a basic property of all matter, and it reflects the rate of heat transferred in response to a given temperature difference. Metals have a very high thermal conductivity, such that they are able to carry away heat very efficiently due to temperature differences with other objects (for example a wet tongue at 98.6°F and a metal lamppost below 32°F in winter) while wood is a pretty poor thermal conductor. Making reasonable assumptions that the Earth was a sphere of molten rock at 7000°F when it initially formed, and knowing the thermal conductivity of rock, Lord Kelvin was able to determine how long it would take the Earth to cool to its present temperature. His conclusion that the Earth was at least ten times too young to provide sufficient time for evolution's effects was considered to be a near-fatal flaw in Darwin's arguments. Kelvin's understanding of thermodynamics was so highly regarded that the Absolute temperature scale mentioned in the previous chapter, in which zero kinetic energy is recorded as zero degrees Absolute, is named in his honor (and is now referred to as "degrees Kelvin"). There was nothing wrong with his calculation.

While Darwin could not refute Kelvin's result, he remained convinced of the validity of the theory of evolution, as it was able to account for too many biological phenomena to be completely wrong, despite Kelvin's objection. Darwin passed away in 1882. A few years later, radioactivity was discovered, and it was quickly realized that there was an additional internal heat source within the Earth that Kelvin had not taken into account, simply because he (like the rest of the world) was unaware of its existence. This extra heat within the Earth would lengthen the time needed for the planet to cool to its present temperature. When Kelvin redid his calculation in 1905, now incorporating the energy provided by radioactive decay, he arrived at a minimum estimate of the Earth's age of several hundred million years. The current determination of the age of the Earth is 4.5 billion years, easily old enough to provide a landscape for evolution to operate. Darwin went to his grave not knowing that Kelvin was mistaken, yet nevertheless maintained his belief in the correctness of his theory of evolution.

There are critics of evolutionary theory today who point out particular biological phenomena that the theory cannot currently explain, but this does not necessarily invalidate a scientific theory. For example, the motion of three masses interacting through their mutual gravitational attraction turns out to be so complicated as to defy analytical calculation, but this does not indicate that the theory of gravity is wrong. There are always gaps in our knowledge and many things we do not presently understand, but the only way we will change this situation is by critical thinking and experimental testing of evidence. If you find the scientific method lacking in one aspect of science, then honesty would indicate that you should refrain from using its results in all other parts of your life. Which will certainly save you some money on doctor and electricity bills.

14

HOW THE MONSTROUS MENACE OF THE MYSTERIOUS MELTER MAKES DINNER PREPARATION A BREEZE—

PHASE TRANSITIONS

NOT EVERY SUPERHERO possesses powers and abilities far beyond those of mortal men. Some, such as Batman and Wildcat, bravely face down supervillains armed with nothing more than a good right hook and the courage to appear in public wearing their underwear on the outside of their clothes. Of course, Batman would try to even the odds somewhat by using his analytical brain, as highly trained as his body, to produce a fabulous array of crime-fighting weapons that he stored within his utility belt. Over at Marvel Comics, the engineer as superhero reached its apogee in *Tales of Suspense* # 39, featuring the debut of the invincible Iron Man. When electrical engineering genius and weapons manufacturer Tony Stark dons his flexible suit of red-and-golden armor, he has the strength of a hundred men, is able to fly using jets built into the soles of his boots, and can fire concussive "repulsor rays" from the palms of his gloves.

We will have much, much more to say about Tony and his golden avenger alter-ego when we get to chapter 23, devoted to a discussion of solid-state physics. Right now I want to consider one of the charter members of Iron Man's rogue's gallery of superpowered villains, who would bedevil him time and again. This villain was one of the first to actually strike fear into Tony's shrapnel-damaged heart.* For if you are wearing a suit of iron, and your only

---

* The "shrapnel" reference will make sense in chapter 23, when we discuss in depth Iron Man's origin. Stay tuned, True Believer!

superpower stems from your suit, then one of your worst nightmares would involve a villain possessing a "melting ray," capable of dissolving iron like hot butter on a stovetop. Unfortunately for Iron Man, Bruno Horgan possessed just such a melting gun, and as the costumed criminal the Melter, he was only too happy to use it. When the Melter first appeared back in 1963, the notion of a melting ray seemed suitable only for comic books. As we'll now discuss, science and engineering have advanced to the point where such devices are commonplace. You probably have one in your home right now (you no doubt refer to it as a "microwave oven").

Before we can answer why solids melt when they become very hot, we need to address a more basic question: Why do atoms combine to become solids in the first place? It all comes down to energy and entropy. Under certain circumstances, two atoms may have a lower total energy when they are close enough to each other that their electrons' "orbits" overlap. When this happens, a chemical bond forms between the two atoms. This lowering of energy is not always very significant. If the two atoms are moving very fast when they are brought together, then their individual kinetic energies will be much greater than any lowering of energy resulting from the formation of a chemical union, and they will bounce off each other and no chemical bond will have formed. It's easier to hook up a trailer to a towing hitch on the back of a truck if you slowly back onto the hitch, rather than smash into it at 100 mph. The slower the atoms are moving, the greater the chance that the resulting lowering of energy will prevail when they overlap, and they will remain linked.

What is true for two atoms will also hold for two hundred, or two trillion trillion, atoms. As the temperature of a gas is lowered, the average kinetic energy of each atom decreases and the greater the chance that the atoms, when they collide, will condense into a new phase of matter—a liquid. Adding thermal energy—heat—to the liquid reverses the process, and the fluid will boil and return to the vapor phase. Similarly, upon lowering the temperature of a liquid, a point is reached at which the atoms cannot glide past one another, and they become locked into a rigid, solid network. If I squeeze on the collection of atoms, I force them to remain closer to one another than they would ordinarily prefer, changing the temperature at which the phase transition occurs.

What determines the exact temperature and pressure at which a phase transition takes place depends on the details of how the individual atoms link up when their electronic clouds overlap. To determine the temperature at which a phase transition such as melting or boiling occurs, we must do more than simply count up the energy needed to break each chemical bond that holds a solid or liquid together. We also have to take into account the large change in the randomness of the atoms—that is, their entropy. For a given internal energy, systems tend to increase their entropy, because all other things being equal in general there are more ways to be in disordered configurations than in neat, ordered piles. The competition between lowering energy and increasing entropy leads to a fascinating collective phenomenon in which all of the atoms in a solid decide to melt at the same temperature. By the way, the bubbling we associate with boiling water in a pot arises from small irregularities in heating at the bottom of a typical saucepan. Individual points on the pan's bottom will be hotter than neighboring regions, and the liquid-to-vapor transition occurs first at these locations. The underwater vapor forms a buoyant bubble that rises to the surface. In an extremely uniform and clean container atop a homogeneous heat source (such as a hot plate), the level of pure water in the container will smoothly drop without any detectable bubbling, as the phase transition occurs at the same time at all points on the container's bottom.

To initiate the melting process, we must add energy to the solid. We can do this the slow, conventional way, by placing the solid in an oven, or the quick way, à la Bruno Horgan and his melting ray. In a conventional oven, the heating elements—either gas flame jets or electrical coils—cause the average temperature inside the oven to rise. A solid placed in the oven, such as a nice roast, will come to the temperature of the oven as air molecules collide with the walls of the oven, pick up some extra kinetic energy, and then make their way to the roast. Striking the surface of the roast, these fast-moving air molecules transfer their energy to the meat. With a conduction oven one must wait for the hot air molecules to randomly make their way from the hot walls to the cooler roast, while with a convection oven a fan generates circulation cells from hot to cold and back again (as in our discussion about Storm of the X-Men in the previous chapter). In either case the surface of the roast warms up first, and one must wait, sometimes several hours, for the center of the meat to reach a higher temperature. As the internal temperature of the meat increases, the atoms shake more and more violently about their equilibrium positions. At a given temperature the shaking of the connecting fibers and deposits of fat in the roast is sufficiently pronounced that these fibers undergo a phase transition and melt.* Since these were the tough, stringy tissues binding the muscle cells in the roast, melting them makes the meat more tender and easier to eat. This is the same principle utilized by the Flash when escaping from the solid blocks of ice in which Captain Cold would routinely entomb him.† One can partially dissolve these fibers chemically, using a lemon juice or vinegar-based marinade, but here again one depends on the chemicals to slowly diffuse into the center of the roast, as in a conventional oven.

If you're in a rush but unable to vibrate at super-speed, there is another technique you can use. This involves grabbing hold of every single atom in the solid at the same time and shaking them back and forth very rapidly, using internal friction to cook all parts of the roast simultaneously. This is what a microwave oven and the Melter's deadly ray gun does.

Every atom in a solid is electrically neutral, with exactly as many positively charged protons in the nucleus as there are negatively charged electrons swarming about it. But the electrons are not always distributed around the nucleus in a perfectly symmetrical manner. Due to the vagaries of the probability clouds and the nature of the chemical bonds holding the atoms together, sometimes one side of the atom may have more electric charge than the other. In this case the atom will be a bit more negative on one side and a bit more positive on the other, just as a bar magnet will

---

* While the thermally driven chemical and structural changes during cooking can be quite complex, for our purposes the key step will be approximated as a melting transition.

† By vibrating back and forth at high speed, the Flash imparts kinetic energy to the ice crystals surrounding him. A vibration rate back and forth of 100,000 times per second, even if he could only flex back and forth half an inch, would correspond to a total kinetic energy of $(1/2)\,mv^2$ of 35 million kg-meter^2/sec^2. The melting transition of ice to water requires the addition of 336 kg-meter^2/sec^2 per gram of ice at 32°F. With his excess kinetic energy, the Flash is able to melt a hundred kilograms of ice around him, freeing him to return Captain Cold to the Central City Jail.

have one end with a North magnetic pole and the other end will be the South magnetic pole. This charge imbalance is not very great, but it gives an applied electric field something to grab on to. Even molecules with perfectly symmetric charge distributions can be polarized by an external electric field.

If a large enough electric field is applied across the solid, the imbalanced atoms line up with the field, just as the needle of a compass will rotate and point in the direction of an external magnetic field. If I now suddenly reverse the direction of the electric field, all the atoms will flip 180 degrees and point in the opposite direction. Changing the electric field back to its original orientation, the atoms will have to rotate once again. If I flip the direction of the electric field back and forth several billion times a second, the atoms are going to do some serious rotating. This energy of vibration will very quickly raise the average internal energy of each atom in the material and, in so doing, raise its temperature. As the external electric field penetrates deeper within the material (with a few exceptions) more atoms will move back and forth due to the oscillating electric field at the same time, not just those on the surface. This process is many times more efficient than waiting for the transfer of kinetic energy by the impact of hot air molecules. The frequency of oscillation of the alternating electric field is in the microwave portion of the electromagnetic spectrum, hence this type of cooking device is called a microwave oven.

Microwave emitters (called magnetrons) were first created for radar applications during World War II. The cooking benefits of such a device were noted in 1945 when engineer Percy L. Spencer, studying the range of microwave energy emitted from a magnetron, noted that the candy bar he was storing in his pants pocket had melted. A follow-up experiment with popcorn confirmed the non-military usefulness of this device.

The easier it is for the atoms in an object to move back and forth and rotate with the oscillating electric field, the quicker the temperature of the object will rise. This is why liquids heat up faster than solids in a microwave. You can dig a deep hole in a large chunk of ice and fill it with water. Placing the "ice cup" filled with water in a microwave oven enables you to boil the water while the outside of the ice cup remains cold and solid. Don't leave the ice cup in the microwave for too long, however, as it will also melt due to the ministrations of the alternating electric field, and in much less time than it would take in a conventional thermal oven.

From the descriptions given in the pages of *Tales of Suspense* and *Iron Man* comics, can we infer that the Melter's weapon used the same principle underlying microwave ovens? Yes and no. Bruno Horgan first appeared in *Tales of Suspense* # 47 as an industrialist competitor of Tony Stark's and was embittered when he lost a government contract to build tanks for the U.S. Army once the military discovered that Horgan was using "inferior materials." Stark's company then won the Army contract, despite the fact that there was an apparent conflict of interest in the information presented, as the report describing Horgan's use of inferior components was written by Tony Stark himself. Later, one of Horgan's laboratory testing devices, built with inferior parts, goes haywire while he is examining it and emits an energy beam that melts any iron it strikes.* When Horgan realizes that the "inspection beam" he has created is actually a melting ray, he redesigns the device into a compact, portable unit and, donning a hideous blue-and-gray costume (sadly reinforcing various stereotypes concerning the fashion sense of engineers), he decides to destroy his enemies and make himself supreme (sadly reinforcing various stereotypes concerning the ethical sense of modern industrialists). After initial success against both Stark Industries and Iron Man, Horgan is dumbfounded when he discovers, at the story's conclusion, that his ray is no longer effective against the golden avenger. The Melter is unaware that Tony Stark has surmised his weapon's weakness: It only works on iron! Creating a suit of "burnished aluminum" that appears indistinguishable from his regular suit of armor, Tony is able to fight the Melter to a standstill, and it is only the accidental melting of an iron drain above a sewer system that enables Horgan to escape to fight another day.

From this story we must conclude that Horgan's melting ray is *not* a portable microwave device. A microwave oven's oscillating electric field grabs hold of *any* atom, while Horgan's weapon works on iron (which contains 26 electrons) but not on aluminum

---

* I cannot stress enough that nearly without exception one cannot randomly combine a collection of circuitry and power supplies into an object and "accidentally discover" that it is a fully functioning death ray (I speak here from bitter experience).

(with 13 electrons). Later on (in *Tales of Suspense* # 90), Horgan's melting ray gun would become even more specific, with dial settings for stone, metal, wood, and flesh (yeesh!). This specificity came to Tony's rescue when, while in civilian clothes, he was shot by Horgan with this weapon, yet was unhurt. Bruno Horgan did not know that Tony Stark was also Iron Man, and was therefore unaware that Stark always wore his metal chest plate underneath his shirt (in order to keep the shrapnel near his heart at bay—see chapter 23) and thus had the gun set to "Flesh" when "Metal" would have been the correct setting.

Now, it is certainly true that when two atoms form a chemical bond, the lowering of energy is unique to the particular atoms participating. Thus, every chemical bond has its own energy signature, and it is, in principle, possible to design a microwave-type weapon that would be tuned to the chemical bonds in stone and not those in metal. Similarly, tuning the resonant frequency to water would make the beam effective against people and not inanimate objects. Such a microwave-based "heat ray" that induces extreme pain, similar to a second-degree burn, while the beam is incident on a person, has in fact recently been developed. The motivation for the development of such a weapon is for use in crowd-control situations, as the heat beam induces a group of people to disperse from a given location in order to avoid burning pain. However, regardless of how the frequency of the oscillating electric field is tuned, the bonds between iron atoms and aluminum atoms in their respective metals are much too similar for a weapon designed for melting iron to not also melt aluminum.

Of course, all this discussion about the Melter and Tony Stark's metallic union suit raises a question that has long plagued modern man: If we can put a man on the moon, why can't we put metal in a microwave? The answer is that we *can* put metal in a microwave, but the free electrons in the metal may cause some serious problems. Metals have high thermal conductivities, and can cause fires when in contact with paper in a microwave oven. Applying an external electric field to these electrons that are able to roam over the entire volume of the metal does more than just push them back and forth as it does the fixed atoms.

Any metal in a microwave is an isolated object, and there is no place for these pushed electrons to go—hence, they can build up at one end of the metal. If there are sharp points or edges, this pileup of electrons can cause a new, large electric field to be created inside the metallic object. If this electric field becomes larger than 12,000 Volts per centimeter, it can cause a spark, as the air is no longer able to insulate the high voltage metal from the wall of the oven, and tiny lightning arcs will emanate. Depending on the metal's curvature, the electric field induced may be less than the critical discharge level, while a sharp corner on a foil-covered stick of butter can be sufficient to create a spark that permanently scars the internal surface of the oven. (Personal note to my wife: Sorry, honey.)

15

ELECTRO'S CLINGING WAYS—

ELECTROSTATICS

UP TILL NOW, we have primarily focused on how forces change the motion of objects, and the force that has concerned us almost exclusively has been gravity. Whether slowing Superman down in his leaps or speeding Gwen Stacy in her fatal plunge, it is gravity that has been invoked when a force **F** is needed in Newton's second law $F = ma$. But there are forces other than gravity in this—and the comic-book—universe.

Physicists have discovered that only *four* basic forces in nature are both necessary and sufficient to account for the wide range of complex physical phenomena we observe. These forces are: (1) Gravity, (2) Electromagnetism, and the unimaginatively named (3) Strong and (4) Weak forces.* The latter two only operate inside atomic nuclei. The Strong force binds protons and neutrons together in close proximity within the atomic nucleus, and without it the positively charged protons would repel each other and no stable elements other than hydrogen could exist. The Weak force is responsible for some forms of radioactivity (such as the nuclear decays that led physicists to suggest the existence of neutrinos, as mentioned in chapter 11), and without it few superheroes or supervillains would exist. Nearly every force we encounter in our everyday dealings, aside from gravity, is electrostatic in nature. The forces generated by our muscles, the force the chair exerts on

---

* These four forces, in the form of the villains Graviton, Zzzax, Quantum, and Halflife, can be quite formidable, as the West Coast Avengers found in "The Unified Field Theory."

the seat of your pants to keep you from falling to the floor, the force exerted by the hot gases in your car engine's cylinder that lead to locomotion, all these and many others are, in the final analysis, electrical. It is thus time for us to consider the twin forces of Electricity and Magnetism, which we will find are really just one single force properly termed "electromagnetism."

Very few superheroes have powers that are electromagnetic in origin. Two of the earliest Silver Age comic-book characters whose powers do utilize electricity and magnetism are Lightning Lad and Cosmic Boy. These heroes are from the future, and they first appeared in *Adventure Comics* # 247 (April 1958) when they, along with Saturn Girl, traveled back in time in order to recruit Superboy into the Legion of Super-Heroes. Lightning Lad is able to create and discharge electrical bolts from his hands, while Cosmic Boy can control magnetic objects. The third founding member, Saturn Girl, possessed the superpower of mental telepathy, which we will argue later is intimately connected with electromagnetic wave propagation. Consequently the three founders of the Legion are direct manifestations of electricity and magnetism theory in action.

The Legion hailed from the year 2958 (current stories take place in 3005) and was comprised of teenagers from different planets who each had a unique superpower. The concept of a club of teenaged superheroes in the future proved very popular with comic readers, and the Legion of Super-Heroes became a regular backup feature in *Adventure Comics* and eventually squeezed Superboy out of his own comic. The Legion membership grew over time, and currently boasts more than thirty heroes. All of the fundamental forces of nature, as well as several basic symmetries of physical laws, were pressed into service as the writers of Legion stories strained to develop a superpower for each hero. Legionnaires include Star Boy who could make objects heavier while Light Lass could make them lighter, Element Lad who was able to transmute one element into another (implying control over nuclear forces), and Colossal Boy who could grow to great heights while Shrinking Violet could miniaturize herself. Ferro Lad could transform himself into some sort of organic iron (an early, teen version of Colossus of the X-Men); as a kid I was profoundly shaken when he nobly sacrificed himself in order to destroy the Sun-Eater in *Adventure* # 353.

While only a few heroes draw upon Electricity and/or Magnetism as the source of their superpowers, supervillains frequently employ these fundamental forces of nature as they seek either financial gain or world domination (and occasionally both). In particular, in the next few chapters we'll focus in turn on two such evildoers, Electro and Magneto (and no fair peeking to see which villain is associated with which force, electricity or magnetism).

STATIC ELECTRICITY—NATURE'S MOST POWERFUL FORCE!

Max Dillon was a highly skilled but self-centered electrical utility worker. When a coworker was trapped atop a high-tension line, Max was cavalier about his fate until his foreman offered Dillon a $100 reward (in 1963 dollars, worth about 600 today) for rescuing him. Freeing the unconscious colleague and lowering him to the ground with a cable, Dillon then receives an unanticipated bonus when he is struck by lightning while grasping the high-tension lines. Just as in the case of Barry Allen (the Flash), not only did Dillon not die or suffer any burns or neurological damage from this traumatic event, but he in fact gained the ability to store electrical energy that he could discharge at will in the form of lightning bolts.* Dillon's accident, presented in *Amazing Spider-Man* # 9 may have changed his body but it left his antisocial attitudes intact. Realizing that he now possessed fearsome electrical powers, he designed a garish green-and-yellow disguise, with a bright yellow lightning-bolt-themed mask, and embarked on a life of crime as Electro, as shown in fig. 22. Personally, if I gained mastery over such a powerful force of nature, I don't think this would necessarily be the costume that I'd choose to wear in public. Perhaps if Max Dillon had not been such a rat, his friends might have gently provided some better fashion advice. But it is exactly such a pattern of bad choices that frequently leads these superpowered miscreants to a life of crime.

Dillon found that his body could store electric charge that enabled him to hurl lethal bolts of electricity. Stories featuring Electro would frequently show him charging up his body at some

---

* Young readers during the Silver Age could be forgiven if they reached the conclusion that being struck by lightning, preferably in conjunction with some other hazardous activity, was one of the best things that could happen to them, second only to being exposed to massive doses of radiation.

Fig. 22. *A scene from* Amazing Spider-Man *# 9, where the supervillain Electro simultaneously demonstrates an advanced concept in electromagnetism and a significantly less sophisticated fashion sense.*

"abandoned" power station, standing between two transformer towers and letting the electric current flow through his body. (The fact that New York City had fully functioning power stations lying dormant throughout town, available for supervillains' use, surely accounts at least in part for the high Consolidated Edison utility bills city residents must now pay, not to mention recent blackouts.) Fully charged, Dillon could project lightning bolts from his hands, though sometimes he would discharge through other parts of his body. Once his stored charge was depleted, he was basically powerless until receiving another charge. Essentially the freak accident on the power line turned Max Dillon into a walking rechargeable taser gun.

What does it mean to have "electrical powers," such that one could hurl electrical bolts at the police and costumed superheroes? Anyone who has shuffled his feet through shag carpeting on a dry winter day and then touched a metal doorknob has verified that matter is composed of electrically charged elements. Unlike the mass of an object, which is always positive, electrical charge comes in two varieties and is arbitrarily labeled "positive" and "negative." The expression "opposites attract" may or may not be a reliable guide in affairs of the heart, but it does accurately summarize the nature of the force between positively and negatively charged objects. Two objects with opposite charges will be pulled toward each other by an attractive force. Similarly, two objects that are electrically charged with the same polarity, either both positive or both negative, will repel each other. When a shipping box picks up an excess electrical charge due to random frictional contacts, this charge can be transferred to the foam packing peanuts inside the box. The fact that all of the foam bits have the same charge accounts for these lightweight peanuts repelling one another and flying into the air when the box is opened. The negatively charged electrons in an atom are attracted toward the positively charged protons in the nucleus by the attractive electrostatic force. The more protons there are, the larger the positive charge and the greater is the force pulling the electron into the nucleus. However, the more electrons in an atom, the greater their mutual repulsion. These two forces—the attraction by the nucleus and the repulsion by the other electrons—tend to roughly cancel out, which is why a uranium atom with 92 electrons and an equal number of positively charged protons in its nucleus is approximately the same size as a carbon atom, with 6 electrons and 6 nuclear protons.

The attractive force between two oppositely charged objects, or repulsive force for two objects with the same charge, has, remarkably enough, the same mathematical form as Newton's law of gravitational attraction described in chapter 2. That is, the force between two objects that have electrical charges **charge 1** and **charge 2** is given by the equation

$$\text{FORCE} = K\ [(\text{CHARGE 1}) \times (\text{CHARGE 2})] / (\text{DISTANCE})^2$$

This expression, attributed to the French seventeenth-century scientist Charles Coulomb, is nearly identical to Newton's gravitational expression, except that instead of the charge of two objects we multiplied their masses, and the constant wasn't called "**k**" but "**G**." Recall from chapter 2 that Newton's law of gravity described the force between two masses **mass 1** and **mass 2** by the expression

$$\text{FORCE} = G\ [(\text{MASS 1}) \times (\text{MASS 2})] / (\text{DISTANCE})^2$$

Mathematically these two expressions for Force are equivalent when "mass" is replaced by "charge" and the constant **G** is renamed as the new constant **k**. Because electric charge is not the same quantity as mass, the units of the constant **k** are different from the units of the constant **G** in order for both equations to have the units of a force.

More important than **k** having different units from **G** is the fact that the magnitude of **k** is very much larger than the magnitude of **G**. Consider a single proton in the nucleus of a hydrogen atom, orbited by a single electron a certain distance away. The attractive force of gravity pulls the electron in toward the proton, and there is an additional attractive force since the positively charged proton is pulled toward the negatively charged electron. The magnitude of the charge on the proton is exactly the same as that of the electron, where the charge of the proton is labeled positive by convention and that of the electron is considered negative. While they may have equal but opposite charges, the mass of the proton is nearly two thousand times larger than that of the electron, so one might think that gravity would dominate over electrostatics. However, the constant **k** in Coulomb's expression is so much larger than the gravitational constant **G** that the force of electrostatic attraction is ten thousand trillion trillion trillion (a one followed by forty zeros) times stronger than the gravitational attraction. On the atomic scale, gravity is irrelevant, and matter is held together by electrostatics. There'd be no molecules, no chemistry and, hence, no life without static cling.

If gravity is so much weaker than electrostatics, why does gravity matter so much for planets and people? Because it is always attractive. Two masses, no matter how big or small, will always be pulled toward each other due to gravity. While there *is* such a thing as antimatter, it has a positive mass, and therefore has a normal gravitational attraction with other matter. As far as anyone has been able to experimentally determine, there is only one type of mass with one type of positive gravitational attraction. Certain puzzling astronomical observations have recently been interpreted as indicating the presence of some sort of "antigravity" associated with a mysterious quantity termed "dark energy." However, this explanation is somewhat controversial, and at the time of this writing scientists don't have the foggiest idea what dark energy is (which is why, though we are well past the year 2000 we have yet to manufacture the flying cars that were promised by this

date in science-fiction stories and comic books of the 1950s and
1960s).

The situation with electricity is very different. The fact that
electrical charges come in two different types—positive and negative—introduces the ability to screen out electric fields. An electron orbiting a proton feels an attractive pull. A second electron
brought near this arrangement is pulled toward the proton but is
pushed away from the first electron. Until the second electron
comes very close to the proton, the sum of the pull and push can-
cels out, and there is no net force on the second electron. If we could
as easily screen out gravitational attraction, then levitating devices
such as the evil Wizard's antigravity discs would be commonplace.
Regardless of whether it has a positive, negative, or no (neutral)
electrical charge, all matter has a positive mass and feels an attractive gravitational pull from other matter. In this way, gravity always
wins out eventually and pulls bodies together—even those that are
electrostatically neutral.

But make no mistake, electrostatics is the stronger force. Consider the form of Coulomb's force equation stated earlier. If you
had only 10 percent more negative charge than positive charge on
your body, then the force of electrostatic repulsion would be large
enough for you to lift an office building that had a similar 10 percent excess negative charge. On the other hand, while the mass of
the office building is much greater than your own, you are not grav-
itationally bound to the building, despite the occasional dictates of
the workplace and your boss.

As he runs, the Flash should pick up an enormous static charge
due to the friction between his boots and the ground that is necessary in order for him to run, just as happens when we rub our feet
along a carpet in the winter. The friction of our feet rubbing against
the carpet, which is a violent process on the atomic scale, results in
the transfer of electrons, which spreads over to our entire bodies.
These excess charges repel each other and don't want to stay on you.
When you approach the doorknob, a path for the charges back to the
Earth (which is able to take up a few more or less electrons without
bother) becomes available. If the charge is large enough, the electrons will jump through the air, the same way a lightning bolt allows
the excess charge in a thundercloud to discharge to the ground.
When driving, your car will frequently pick up excess charge due to

the friction between the tires and the road, which you can remove by touching the metal doorframe once the car has stopped. The discharge is painful for two reasons: The surface area of your fingers is very small, so the current per area is large (better to touch the car frame or the doorknob with your elbow or drape your entire body over the metal object. The arch looks you'll receive will be a small price for the reduced pain), and your fingertips have more nerve endings, so they are more sensitive to the current.

This friction-induced electrostatic buildup (technically referred to as "contact electrification") was recently acknowledged in *Flash* # 208. The Flash had just finished saving the citizens of Keystone City,* yet again, from an attack by a subset of his Rogue's Gallery, and was being thanked by a group of bystanders. In addition to requests for his autograph, one person patted the Flash's shoulder and, receiving a shock, noted, "Hey, check it out! His uniform is covered in static electricity!" This excess charge should in general discharge to the first metallic object near the Flash that was connected to electrical ground as soon as he stopped running. The fact that this contact electrification was only noticed in 2004 (and not in the previous fifty years of *Flash* comics) suggests that during the majority of his crime-fighting career the Flash, in addition to possessing an ability to ignore air resistance and punishing accelerations, was similarly immune to electrostatic buildup.

Returning to Electro, his electrical powers no doubt stem from the fact that he is able to store very large quantities of a net electrical charge, either all positive or all negative, within his body. He can then discharge himself at will, in a similar fashion as the spark that leaps from your fingertip to the brass doorknob mentioned earlier. This is consistent with the fact that Electro needs to charge himself up before employing his powers, and if he lets loose with too many electrical bolts, he is, in essence, depleted and susceptible to a good right hook.

* * *

Roughly sixty years ago, a Swiss engineer's hiking frustration led to a technological innovation. George de Mestral's investigations

---

* This being the Wally West Flash, who based his operations out of Keystone, the Twin City to Central City, home of the Barry Allen Flash.

into why burrs clung so tenaciously to his pants resulted in the invention of a fastener consisting of millions of tiny hooks and loops, and gave the world Velcro. More recently, Robert Full, Keller Autumn, and coworkers have discovered that the gecko's ability to climb up smooth walls and ceilings can be traced to millions of microscopic hairs on the lizard's toes called "setae." But without miniature hooks in the walls or ceiling, what holds the fibers and the attached gecko in place? Static cling!

The fibers in a gecko's feet are electrically neutral, but the lizard does not need to shuffle across a shag carpet to cling to a wall, because he makes use of fluctuations of charge in his setae. The electrons in the fibers in the gecko's toes are constantly zipping around. Sometimes a few more electrons are on one side of the fiber, making that side slightly negatively charged, while other times a few less electrons are on that side, making it slightly positively charged. If the side of the fiber closer to the wall is, just for a moment, slightly negatively charged, then it will induce a slightly positive charge in the wall, and an attractive force between the fiber and the wall will result. You would expect that this force (known as the van der Waals force) is very weak, and you would be right. Which is why the gecko has *millions* of these fibers in each toe, so that the total attractive force can be large enough to support his weight.

Or possibly even Peter Parker's weight. Marvel's writers have suggested that Spidey's wall-crawling ability is electrostatic in nature—the 2002 *Spider-Man* film included a scene showing scores of microscopic barbed fibers sprouting from Peter Parker's fingers once he had gained his spider-powers. Turns out that both the comic and the movie are on to something. Recently a report from the University of Manchester in England described the development of "gecko tape," consisting of millions of tiny fibers (the length of each fiber is fifty times shorter than the width of a human hair) that can provide a strong enough attraction to support a Spider-Man action figure from just one palm. A tape that makes use of the force arising from fluctuating charges can be, in principle, instantly used and reused, unlike a single application adhesive that requires a curing time. The fibers on the tape must be very small in order to maximize the ratio of surface area to volume, since only the fluctuating excess charges on the surface of the fiber contribute to the attractive force. In order to provide enough force to support a grown person's weight, the density of microfibers must be very high, to compensate for the extremely weak force of each fiber. Whether these engineering challenges can be resolved remains to be seen. But if "gecko tape" ever becomes as common as Velcro, I for one will never wait for the elevator again!

SUPERMAN SCHOOLS SPIDER-MAN—

ELECTRICAL CURRENTS

LET'S NOW TAKE a closer look at those electric bolts emanating from Electro's hands. A large enough positive charge can pull electrons from very far away, even through miles of copper wire. A fancy term for the pull exerted on electrons as they move through a wire is "voltage." Electrons are negatively charged so that a positive voltage pulls them in one direction while a negative voltage repels the electrons in the opposite direction. The current is just another way of expressing the number of electrons moving past a given point in the wire per second.

Imagine a garden hose connected to an outdoor faucet. In this case the voltage plays the role of the water pressure that pushes the water through the hose. The amount of water that comes out the end in a given time period is the current. The resistance of the hose arises both from minor blockages as well as small holes along its length, from which some water can escape before making it to the end. The more defects in the hose, the greater the water pressure needed to maintain the same water flow (current) from the end of the hose. However, just like a faucet being turned on in a sink—with water flowing without a hose connecting the faucet to the drain—an electrical current can be pushed by a large enough voltage even in the absence of a wire. This is what happens when a spark jumps from your fingertip to the doorknob or from a cloud to the ground during a lightning strike. The greater the distance, the bigger the force needed to pull the charges. This is a consequence of the Coulomb electrostatic force expression becoming smaller by the square of the separation of the charges. A long garden hose, with various imperfections and holes, will have more resistance to water flowing through it than a similar short segment of hose. This is why you don't receive a static shock until your fingers are very close to and just about to touch the doorknob: Air is a pretty good electrical insulator, and it takes an electric field of more than 12,000 volts per centimeter before the pull on the electrical charges is sufficient to make them jump the gap. Which is why when it happens, it stings. And why you most definitely do not want to be zapped by Electro's massive discharges.

When you turn on the water in the kitchen sink, the water flows from the tap to the drain. It does not flow from the faucet to the ceiling,* under ordinary circumstances. Why not? For the water analogy, the reason is obvious: There is a downward pull of gravity on the water, directing its flow. For electrical charge, the direction the current flows is determined by the location of the "drain." Electrical charge can't flow if there is no place for it to go. Actually, this is true for our water analogy as well. Want to know how you can overturn a glass of water filled to the brim and manage to not spill a drop? Do it when the glass is underwater in a swimming pool! If there is no place for the water in the glass to go, it will stay inside the container (provided you ignore random collisions between the water molecules in the pool and those at the top of the glass that will cause those molecules to switch locations).

The same is true for electricity. Regardless of the magnitude of the net electrical charge an object possesses, it will not discharge if every other object surrounding it has exactly the same charge. Technically the voltage that pulls or pushes electrical charges around is a measure of the "potential difference," which is defined as the difference in potential energy of the charges on one object relative to another. This is what makes Electro so dangerous (in addition to his daring fashion sense): He is able to control his potential difference relative to his surroundings at will, so he can decide when and where he will discharge the excess electrical charge he has stored up.

By applying a voltage across a conductor, I raise the potential energy of the electrons within the conductor, just as when lifting a

---

* Unless you employ the Moe, Larry, and Curly plumbing agency.

brick over my head I raise its potential energy. The brick keeps this extra potential energy until I release it, at which point the potential energy is converted into kinetic energy and the brick speeds up as it falls. But this conversion cannot take place until I let go of the brick. Similarly, the electrons in a wire speed up and increase their kinetic energy in the form of an electrical current, in response to the voltage applied across the wire—but only if the electrons have someplace to go. Just as the raised brick will keep its potential energy indefinitely until I drop it, the electrons cannot speed up in response to an applied voltage if the wire is not electrically connected to anything. Think again of a garden hose connected to a faucet. No matter how much I turn the faucet tap, absolutely no water will flow through the hose if it is completely sealed at the other end. I have to uncap the end of the hose so that the water can drain out before it flows through the hose (a current), in response to the water pressure (voltage) at the faucet. The technical way of expressing this is to say that in order for an electrical current to flow through a wire, it must be grounded.* The Earth, or "ground" is obviously a large object, with many electrical charges; consequently, it can take up extra electrons, or donate electrons to a wire without difficulty. This notion, that in order for a current to flow, it must have someplace to go, is fairly reasonable, but not every superhero seems to have grasped it.

In chapter 1 we mentioned the early exploits of the Man of Steel, as described in *Superman* # 1, before the world at large knew of his existence. In this story Superman sought to learn the identity of the person bankrolling the Washington lobbyist who was bribing a senator with the goal of embroiling our country in the war in Europe. (Recall that this story occurred in 1939.) The secret employer of Alex Greer, "the slickest lobbyist in Washington," turns out to be Emil Norville, the munitions magnate (war being

---

* Strictly speaking, the wire does not have to be connected to "ground," but simply to another point at a lower potential than the starting point. But, ultimately, the terminus of any current must be the ground. Following the water analogy, we can have a flow of water through a hose that is connected from one faucet at one end to *another* faucet at the other end. As long as there is a difference in water pressure between the two faucets, there will be a net flow of water, but for this to keep up indefinitely the second, lower-pressure faucet must be able to eventually discharge its excess water down some drain (or "ground" back to the electrical situation).

good for business, from Norville's point of view). For some reason Greer initially refuses to divulge the name of his employer to this strange man wearing a blue-and-red long-underwear ensemble accessorized with a flowing red cape. In chapter 1 we mentioned that Superman purposely falls from the top of a high building holding Greer, pretending that the fall will kill them both. Prior to this scene, in order to loosen Greer's tongue, Superman picks him up like a sack of potatoes and leaps atop some high-tension lines, as illustrated in fig. 23. Greer protests that they'll be electrocuted, but Superman finds the time to give the lobbyist a physics lesson. Whether this lecture should be considered an additional part of Superman's efforts to psychologically torture and force information from the lobbyist I leave to the reader to decide. "No, we won't," the Man of Steel explains, since after all, "birds sit on telephone wires

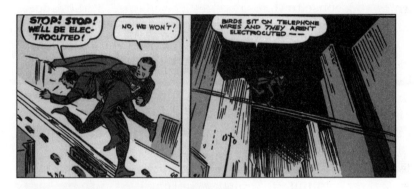

Fig. 23. A scene from Superman # 1, where the defender of truth, justice, and the American Way coerces information from a Washington lobbyist, by giving him a "hands-on" demonstration of the principles of electrical grounding.

and *they* aren't electrocuted—not unless they touch a telephone-pole and are *grounded*!"

Superman is exactly right. It is only when you touch a high-voltage wire and simultaneously grab the telephone pole (or touch another wire at a different voltage), and thereby provide a pathway for the current in the wire to flow to the lower voltage, that you have to worry. In this unfortunate situation the flow of electrons (a current) passes through the conductor—namely, your body—connecting the two points.

Alas, just such a basic understanding of electrical circuitry is lacking in *Amazing Spider-Man* # 9, where Spidey first tangles with Electro. In one scene during their climatic battle, Spider-Man manages to deflect an electrical bolt that Electro has hurled at him by tossing a metal chair over Electro's head. "Anyone with any knowledge of science knows that anything metal can act like a lightning rod," Spider-Man says, lecturing Electro, "as this steel chair is doing!" Actually, Spider-Man's mistaken understanding of how lightning rods function suggests that his allegedly advanced knowledge of science isn't all it's cracked up to be. The electrical bolt is shown arcing away from Spider-Man and chasing after the soaring chair—even though the chair is not electrically connected to anything! Why would Electro's lightning bolt be pulled toward the chair, metal or not, if once it reaches the chair there is nowhere for the electrical current to go?

The situation further degrades (from a physics point of view) in *Amazing Spider-Man* Annual # 1 (Feb. 1964) where Spidey again faces off against Electro and this time, as an extra precaution, he deliberately attaches a wire to his ankle to ensure that he remains electrically grounded at all times! When fighting a supervillain capable of hurling lethal lightning bolts at you, a good solid electrical connection to the ground is exactly what you *don't* want.

The whole point of a lightning rod is not that it's made of metal, but that the lightning will strike the tallest feature on the building (the rod), and the electrical current is then carried from the rod by way of a wire safely to ground, thereby avoiding igniting a fire on the roof of the building. The static shock between your fingertip and the metal doorknob occurs only when your finger is very close to the door, since the shorter the distance the less resistance the arc has to overcome. Similarly, the lightning bolt is trying to minimize the distance and hence resistance on its way to electrical ground. This is why you don't want to stand under a tree during a lightning storm, as you increase the chance that the lightning striking the tall tree will take a detour through your body. When alone in an empty field during a thunderstorm, one should lie flat on the ground in order to decrease the chance of being struck by lightning. If a building's lightning rod is not connected to ground, the electrical current entering the rod will find a higher resistance pathway to ground, through the roof and building, with concomitant damage to the structure.

Just such damage will surely be Spider-Man's fate when he intentionally connects himself to ground, thereby guaranteeing that all of Electro's electrical energy will pass through his body on its way to a lower potential state. Spider-Man's "spider-strength" will enable him to withstand some of the damage of the electrical strike, but grounding himself as he does makes the situation much worse than it needs to be.

It's not clear which of Spider-Man's co-creators, the writer Stan Lee or artist Steve Ditko, should get the blame for these goofs. This ambiguity stems from the "Marvel method" of producing comic books in the 1960s. At Marvel's Devilish Competitors (as Lee would jokingly refer to DC Comics), a comic-book writer would generate a script that detailed not only the captions and dialogue and thought balloons in each panel, but also what the artwork in each panel should look like. An editor would then go over the script, making changes as needed, and pass it along to the artist, who would draw the comic story as described in the script. The artwork would then be inked, lettered, and colored, using the dialogue and captions in the script, and the writer would commonly not see the story again until it was available for sale on newsstands. This system works fine as long as one has enough writers and editors to cover the number of comics produced per month, but at Marvel in the early 1960s, the number of writers and editors was low—namely one: Stan Lee. Lee was both the editor and writer of (in 1965, to pick a particular year) the *Fantastic Four; Spider-Man; The X-Men; The Avengers;* Captain America and Iron Man stories (both in *Tales of Suspense*); Dr. Strange, solo Human Torch stories, and Nick Fury, Agent of

S.H.I.E.L.D.* (in *Strange Tales*); Giant-Man, the Sub-Mariner, and the Incredible Hulk (in *Tales to Astonish*); Daredevil; and *Sgt. Fury and His Howling Commandos* (a World War II comic). If the stories in the Marvel universe had a coherent structure and feel, this was no doubt due to the fact that there was a single creative voice guiding the varied comic books.

With so many stories being created every month, there was simply no way that Lee had the time to craft full scripts for all of these comic books. Meanwhile the artists working for Marvel were freelancers, and would bring in the artwork for one issue, get paid, and then need to pick up instructions for the next issue's story line (if they weren't working, they weren't being paid). I should mention, by the way, that the artists working for Marvel at the time were some of the very best in the business and included such titans as Jack Kirby, Steve Ditko, Don Heck, John Romita, and Gene Colan. These artists were so talented that they were able to continue making a living through the comic-book Dark Ages of the mid-1950s, when the entire industry was on the verge of extinction thanks in part to the "Seduction of the Innocent" brouhaha. Consequently, they were experts in how to tell a story in graphic terms, and did not need a comic-book writer holding their hands with panel-by-panel instructions of what should be drawn on every page.

Stan Lee therefore hit upon a clever solution to the problem of not enough time and too much available talent: Let the artists tell the story. Lee would write up a brief synopsis, varying in length from a few pages to a few paragraphs,† describing what the latest issue's story line would be. In essence, he gave the artists a plot outline of the major points of the story, such as who the villain would be and what his powers were and how he obtained them, as well as how the hero would lose the initial skirmish with the villain, and finally the clever stratagem that would provide the hero's

---

* Supreme Headquarters International Espionage Law-Enforcement Division.

† Or sometime even less. One legend has it that Lee's précis to Jack Kirby for the classic *Fantastic Four* # 48, where the FF encounters the cosmic being Galactus read as follows: "The FF fight God." Kirby, reasoning that such an entity would have a herald to prepare the unlucky planet for the devourer's arrival, created the character the Silver Surfer. Lee first learned of the Surfer when Kirby's artwork for FF # 48 arrived at the Marvel office.

victory by the issue's conclusion. The artists would then go back to their studios and construct a graphic story that followed Lee's synopsis. When the finished artwork would return, Lee would then write the captions and dialogue in each panel, and the comic book would be ready to be sent off to the printer. Consequently both Lee and the artists could legitimately be said to co-write or co-plot any given issue of a comic book created in the Marvel style. Thus it is on both Lee's and Ditko's shoulders that the blame for Spidey's ignorance of basic electrical current theory must be placed.

Lee and Ditko may not have had a firm grasp of the concept of electrical grounding, but they understood that electricity plus water equaled a short circuit. The climax of Spider-Man's battle with Electro in issue # 9 came when Spidey grabs a nearby fire hose, such as used to be commonplace in most professional buildings before the advent of ceiling-based sprinkler systems, and doused Electro with a heavy spray of water. As Spider-Man grabs for the hose and turns open the main pressure valve, he thinks *Say!! What kind of science major am I, anyway? Why didn't I think of this right away??* As he lets Electro have a full blast, he continues, *Water and electricity just don't mix!!!*

Well, as mentioned above, we are beginning to have some doubts about what kind of science major Peter Parker is, but he is certainly correct that water and electricity don't mix. This is because city-water, while technically electrically neutral, contains a high concentration of impurity ions. Ordinary tap water is consequently a pretty good conductor of an electrical current. Electro is at a high potential difference, which is why he is such a lethal threat to superheroes. By dousing him with water, Spider-Man essentially connects a wire between Electro and ground, allowing the large excess charge Dillon has stored to flow out of his body. This is one physics lesson that seems to be well learned in the Marvel universe. When Electro is bested by Daredevil in the second issue of that hero's comic, the police keep him drenched with a water hose (Electro, that is, not Daredevil) in order to safely transport him to the paddy wagon and the station house.

17

HOW ELECTRO BECOMES MAGNETO WHEN HE RUNS—

AMPERE'S LAW

I SUPPOSE THAT WE SHOULD cut Lee and Ditko some slack for their goof described in the previous chapter concerning whether metals conduct an electrical current if they're not connected to ground. The rush to put out a monthly comic book, combined with the need to tell an exciting story, has certainly accounted for more than a few science blunders at both Marvel and DC over the years. As emphasized earlier, these comic-book tales were never meant to double as physics textbooks. It is therefore all the more impressive that in the very same issue of *Spider-Man* that features Electro's debut we see a perfect illustration of a mysterious and fundamental property of electricity. At one point in the story, following a brazen daytime bank robbery, Electro is shown escaping from the authorities by climbing up the side of a building as easily as Spider-Man. The panel is reproduced on p. 169 in fig. 22, where we see one observer exclaim, "Look!! That strangely-garbed man is racing up the side of the building!" A second man on the street picks up the narrative: "He's holding on to the iron beams in the building by means of electric rays—using them like a magnet!! Incredible!"

There are two feelings inspired by this scene. The first is nostalgia for the bygone era when pedestrians would routinely narrate events occurring in front of them, providing exposition for any casual bystander. The other is pleasure at the realization that Electro's climbing this building is actually a physically plausible use of his powers. Utility pole man Max Dillon (Electro) understands, as does the second passerby in the panel, that electric currents do indeed create magnetic fields. This phenomenon, termed the Ampere effect, was first noted by Hans Christian Ørsted (for whom a unit of magnetic field strength is named) and was fully explained by André-Marie Ampère (after whom the unit of electrical current, the Amp, is named). Why does Electro's control of electricity enable him to generate magnetic fields, and wouldn't fairness therefore dictate that Magneto, the mutant master of magnetism, be able to control electric currents at will? The answer to this question reveals a deep symmetry between electricity and magnetism, found in both comic books and the real world.

An electric charge at rest exerts a force on another electric charge. The farther away this second charge is, the weaker the force and, depending on its polarity, the second charge will be either pulled toward or pushed away from the first electric charge. We can therefore say that there is a "zone of force" surrounding this first electric charge. Another way to describe this "zone of force" is to say that around the first charge is an "electric field." A second electric charge brought near the first charge will experience a force, as it is being pushed or pulled by the electric field of the first charge. The strength of the electric field depends on the magnitude of the electric charge at this point and varies with distance away from the first charge—very close to it the force on a second charge is large, while as the separation increases the force decreases as the inverse of the square of the distance (from the Coulomb expression). If the separation of the two charges is doubled, the force goes down by a factor of four, and if the separation triples, the force is only one ninth as large.

There is another field that is created by an electric charge, but only when it is in motion, called a "magnetic field." If an electrical current flowing through a wire is held near a compass needle, the needle will be deflected just as if a magnet were brought near the compass (this was Ørsted's discovery). In fact, two parallel wires carrying electrical current will, depending on the direction of the currents, be either attracted toward each other or repelled away from each other, behaving just as two magnets would when oriented North to South poles (attracting) or South to South (repelling). The magnetic field generated by Electro's "electric rays" does indeed provide an attraction to the magnetic field of the iron beams in the building, enabling him to scale buildings or adhere to passing automobiles (as he did to effect an escape in *Daredevil* # 2).

The force that arises between current-carrying wires is not electrostatic in nature. The wire is electrically neutral before the current flows, with the number of electrons in the atoms making up the wire evenly balanced by the same number of positively charged atomic nuclei. While a current is passing through the wire, the same number of electrons enter at one end as leave at the other. This extra force between the wires when carrying a current is due to the magnetic field they create.

Why is this so? Why does an electric current create a magnetic field, which in all aspects is identical to that of an ordinary magnet? A key clue behind the phenomenon of magnetism is that it involves electric charges in *relative motion*. That is, the charges must be moving relative to each other.

If two electric charges are moving in the same direction at the same speed, then from the point of view of one of the charges, the other charge is stationary. In this case the only force between the two charges, from their point of view, is electrostatic. To someone stationary in the lab there is an extra force associated with the motion, called magnetism. That the magnetic force is connected with the relative motion of the electric charges, suggests that there's a simple four-word explanation for the phenomenon of magnetism: Special Theory of Relativity. To explain how Einstein's theory from 1905 is able to account for magnetic fields will require more than four words, but we'll try to get by without mathematics.

I'll use a nice argument given by Milton A. Rothman in his excellent book, *Discovering the Natural Laws,* that illustrates how relative motion of charges can create a force in a situation where there is no force when the charges are stationary. Think about two very long train tracks lying next to each other, one with a large number of negative charges equally spaced exactly one inch apart, the other with an equal number of positive charges, also one inch apart. We're making this up as we go along, so we'll assume that these rows of positive and negative electrical charges extend for miles and miles, and this way we won't have to worry about running out of charge as they move along the track. We next bring in a test charge—a positive charge, for sake of argument—some distance from these lines of charges. This test charge will feel no net force, as it is pushed away from the line of positive charges as strongly as it is attracted to the negatively charged array. Now the two tracks start moving at the same speed in opposite directions, the negatives to the left and the positives to the right. If the test charge is stationary, then the same number of negative and positive charges per unit length pass by it, and there is still no net force. An extra force develops, however, if the positive test charge moves to the right at the same speed as the positive charges on the track, also moving to the right.

Back in chapter 6, when we were discussing the effects of the Flash's great speeds, we discussed the property of the Special Theory of Relativity that, from the point of view of a stationary observer, the length of the moving object is contracted. From the positive test charge's "point of view," moving along with the same speed and direction as the positive charges, it is stationary compared to this array of positive charges. The test charge therefore sees the positive charges on the track as still being spaced one inch apart. The array of negative charges moving in the opposite direction, on the other hand, will be contracted in length and will therefore be closer than one inch apart to the moving test charge. The electrostatic push and pull on the test charge is now unbalanced, and it will now feel a net attractive force. We give this extra force, that is only present when charges are moving relative to each other, a special name: magnetism. From this argument it is clear that a moving object having no net charge (that is, is electrically neutral) will not feel any extra force, which is consistent with the experimental fact that magnetic fields are only created by positive or negative currents.

Earlier we noted that the friction between the Flash's boots and the ground should transfer static charge to the Scarlet Speedster. Because moving electric charges create magnetic fields, it is puzzling that the Flash, whenever he sprints at super-speed, does not generate an enormous magnetic field that would pull every iron object not nailed down (and quite a few that are) after him in his wake. We will have to chalk both this missing electric charge and corresponding magnetic fields to the efficacy of his "aura" that also enables him to avoid the deleterious effects of air resistance.*

---

* When that extradimensional imp Mopee bestowed super-speed and a protective aura to Barry Allen in *Flash* # 167, he knew what he was doing! Mopee disappeared at the end of the story before Barry Allen could ask him how Wally West (Kid Flash, at the time) acquired his aura. Nowadays, all Flash's auras derive from the Speed Force, which makes as much physics sense as a tenth-dimensional imp.

It is indeed strange that magnetism is explained by invoking the special theory of relativity for moving electrical charges, as it's easy to ignore relativity when the object in question is moving much slower than the speed of light. There's only a very slight error if we neglect relativity even when the object is moving at one tenth the speed of light. Nevertheless, for electrical charges moving much slower than light speed, there is still a relativistic effect through the creation of a magnetic field. The effect is smaller, to be sure. How much smaller? One can show mathematically that the magnitude of the magnetic field created by a moving charge is equal to its electric field divided by the speed of light. The speed of light is a very large number; consequently, for a given electric field, the magnetic field associated with the moving charge will be weak, but it will be there nonetheless. Increasing the magnitude of the electrical current, either by moving more electric charges, or having them move faster, will generate a larger magnetic field.

An understanding of Ampere's law connecting electrical currents and magnetic fields makes possible useful devices such as electromagnets. An electromagnet is a coil of wire wrapped around an iron magnet core. The current flowing through the coils creates a magnetic field, enhancing that from the iron alone. Just such a device is constructed in *Superboy* # 1, when the Teen of Steel stops a gang of thieves who are running around town in a fleet of personal tanks stolen from an army surplus depot guarded by only a few easily overpowered night watchmen. These 1949 versions of Hummers allow the crooks to terrorize the townspeople of Coastville at will. The "Smash and Grab Gang" of crooks (yes, that was really their name) then use these tanks to break into banks and cause mayhem. While he could just as easily fly around and collect all the tanks by hand, as shown in fig. 24, Superboy decides on a more technological approach. "I'll only need that locomotive, a dynamo from that power-house, and a few miles of wire," the teen titan explains to a recent victim of the crime gang. Superboy flies a large electrical dynamo to the empty coal car behind a large locomotive engine, and notes, "This dynamo will give the current I need when it's hooked up! Now for the windings—" In the next panel we see him take "a few seconds to wind these miles of wire" as he loops them around and around the body of the locomotive. On the next page (fig. 25) we see the payoff as Superboy starts up the engine (presumably there is enough coal to get it going) and

Fig. 24. *Superboy demonstrates a practical knowledge of electromagnetic theory, as he constructs a portable electromagnet (from* Superboy # 1)

Fig. 25. *A continuation of the scene from* Superboy # 1, *where the Teen of Steel uses his brains, in addition to his brawn, to capture the* "Smash and Grab Gang."

announces, "I've got the biggest electromagnet ever made, and one that can go places!" The train tracks conveniently pass not only through the center of town, but right by the vandals' tanks. "What happened? We're flying!" one crook cries out as his tank is drawn toward the magnetic locomotive. "It's that locomotive!" says a more informed villain. "It's a magnet drawing our steel tanks!"

This is all perfectly correct, from a physics perspective. The dynamo is the source of the electrical current that passes through the few miles of wire around the engine. The current in the wire creates a magnetic field that projects from the center of the wire loop. If a magnetic material such as a locomotive engine is placed within the loop, it enhances the generated magnetic field. However, why the strong magnetic fields generated by Superboy's homemade electromagnet do not cause the steel wheels of the locomotive to seize up, thereby preventing them from rotating, remains a mystery.

HOW MAGNETO BECOMES ELECTRO WHEN HE RUNS—

MAGNETISM AND FARADAY'S LAW

THE VERY FIRST VILLAIN the X-Men encounter in *X-Men* # 1 is Magneto, the mutant master of magnetism, whose superpower consists of the ability to generate and control magnetic fields. Magneto could hurl missiles at our heroes and deflect magnetic objects, yet would be powerless against a wooden baseball bat. In fact, even some metallic objects are immune to Magneto's power: He is able to pick up an automobile easily enough, but not a silver spoon or gold bracelet. What determines whether some materials are magnetic, even without an electrical current passing through them, and others not? Where does magnetism come from?

The Special Theory of Relativity may be ultimately responsible for the magnetic field created by an electrical current which involves the motion of electrical charges, but what about magnets made of iron? The magnets we use to hold grocery lists to the front of our refrigerators don't seem to have any moving parts, yet they still create magnetic fields. It turns out that relativity is ultimately responsible for the magnetism of a stationary hunk of iron too.

Every proton, electron, and neutron in the universe has a tiny magnetic field associated with it. This field is barely noticeable compared to the Earth's magnetic field, or fields created by electrical current. The electrons orbiting the nucleus can be (very roughly) considered as tiny current loops that generate magnetic fields. But even without this "orbital" effect, there is still an internal magnetic field within atoms. Where does this minuscule intrinsic magnetic field of sub-atomic particles come from? The answer involves quantum mechanics, which we will discuss in the next section. A principle of the Special Theory of Relativity is that space and time should properly be considered as a single entity, called space-time. When this relativistic adjustment is made to the basic equation of quantum physics, the theory predicts that electrons, protons, and neutrons should have a very small internal magnetic field, the magnitude of which agrees precisely with the measured value. The internal magnetic field of elementary particles is understood mathematically only in the relativistic version of quantum mechanics, where time and space are treated on an equal footing within a four-dimensional "space-time." Even for stationary matter, relativity turns out to be crucial for understanding magnetism. So, no Einstein, no relativity, and hence no magnetism. No magnetism, no magnetic iron, and most importantly, no refrigerator magnets! Therefore without relativity there is no way to keep our shopping lists from falling to the floor and lying there, discarded and unread. Absent Einstein's towering achievement in theoretical physics, a slow and lingering death of starvation would await us all.

Normally the small magnetic fields of electrons and atoms like to pair up, just as when you bring two magnets together they orient themselves to align at separate poles. When the magnetic fields inside an atom pair up, there is no net magnetic field associated with the atom, just as an ordinary atom will have no net electric field, because the number of positive protons in the nucleus is balanced by an equal number of negatively charged electrons. Most materials, such as paper and plastic, are not magnetic, and even most metals, such as silver and gold, have all of their magnetic moments paired up.*

If most materials do not have net magnetic fields because their atomic magnetic moments are paired up, then how is Magneto

---

* This point was illustrated in *Atom* # 3, in which the Tiny Titan travels back to the past using a Time Pool discovered by physics professor Alpheus V. Hyatt. By shining a light of "all colors—even the finest shadings" on a small region in space, Prof. Hyatt created a small portal into other times. The opening of this Time Pool is only about six inches in diameter, so the professor dangles a magnet on a fishing line and tries to retrieve magnetic objects from the past (he is apparently unconcerned about what "butterfly effect" any small change in history will have on the present). The Atom, able to shrink small enough to fit through the portal, would occasionally hitch along on the dangling magnet and have adventures in the past, as in the story

able to levitate himself and other people, shown in fig. 26? The physical basis behind this trick is that Magneto is able to generate such a large magnetic field that he essentially polarizes the internal magnetic fields of our atoms, turning us, or any other object, into a magnet.

Before we begin this discussion of magnetic levitation, I first must stress that Magneto does not lift people through his influence on the iron in their blood. Let's leave aside the question of the effect of an nonhomogeneous pressure on the veins and arteries in a person's body, and focus instead on the blood's magnetism. A few metals, such as iron and cobalt, have just the right configuration of non-paired electrons' magnets such that the atom has a net nonzero magnetic field. However, the iron in your blood occurs primarily in the form of hemoglobin, a protein used to collect and transport oxygen and carbon dioxide as you breathe. Hemoglobin is a very large molecule that consists of four large proteins (called globins, and which look like folded worms) bonded together. Each of these proteins contains a large molecule termed a "heme" group, composed of carbon, nitrogen, oxygen, hydrogen, and iron. The iron atoms in the center of each heme molecule are chemically bonded to their neighboring atoms. There's another technical term for an iron atom chemically bonded to oxygen atoms: rust. Rust, as anyone who has dealt with scrap metal can confirm, is weakly magnetic. The common form of rust has three oxygen atoms bonded to two iron atoms (called "hematite") and is non-magnetic, though four oxygen atoms bonded to three iron atoms (termed "magnetite") is magnetic. The magnetic field of the iron in hematite disappears when it combines with oxygen atoms, because the iron and oxygen chemically sharing their electrons pair up the remaining uncanceled electronic magnets in the iron. Depending on whether the hemoglobin has picked up an extra oxygen molecule to bring to the cells, or is carrying a carbon dioxide molecule to be exhaled,

---

involving the change from the Julian to Gregorian calendar mentioned in the Introduction. At the conclusion of the tale in *Atom* # 3, the Atom brings back a gold coin from Arabia in the days of 500 A.D. (the coin has an odd mix of Roman numerals and Arabic writing on it). The Atom holds on to the coin as the magnet is pulled back into the present day, noting, "The professor will be mystified at how this magnet held on to a gold coin!" And indeed Prof. Hyatt was, and properly so.

Fig. 26. *Scene from* X-Men # 6 *(above) and* X-Men #1 *(left), as Magneto threatens or escapes from the mutant X-Man the Angel (the one with the wings), illustrating Magneto's ability to levitate nonmagnetic objects such as a boulder or himself, through the principle of Diamagnetic Levitation.*

the iron can either have an uncanceled magnetic field or not. But at any given moment only a fraction of your blood is even capable of being affected by an external magnetic field.*

---

* In the motion picture *X-Men 2: X-Men United*, Magneto is able to overcome his guard and escape from his plastic cell only after his accomplice Mystique has injected a small quantity of magnetic metal into the guard's bloodstream.

Even when iron is not chemically bonded to oxygen atoms, it is possible that it will be nonmagnetic, if all of the individual atoms are not properly aligned. Ordinarily the atoms inside a piece of iron or cobalt will line up, forming small regions termed "domains" where all the iron atom's magnetic fields point in the same direction. However, entropy considerations lead to the domains pointing in different directions, so their combined magnetic fields cancel out. Heat up a bar of iron so that the atoms have a lot of thermal energy and are free to rotate, and then place it in a strong external magnetic field. The external field induces the majority of the domains to all point in the same direction, so that the piece of iron, when cooled back to room temperature, has a large net magnetic field. If you hit the magnetized iron bar with a hammer or heat it in an oven, you will cause the magnetic domains to reorient themselves randomly, with the effect that the magnet will lose nearly all of its field strength. Some flexible, credit card–size refrigerator magnets have their magnetic domains aligned in little strips along their length. Rather than have all of their domains point in the same direction, it is easier to line them up so that one strip has its north pole pointing toward the refrigerator, while the adjacent strip has its north pole heading away from the fridge, and so on.*

Materials that form magnetic domains with neighboring atoms' magnetic fields pointing in the same direction are called "ferromagnetic" (so named after iron, the most famous example). Many atoms in solids have a very weak magnetic interaction with their neighbors, so if placed in a strong external magnetic field, they will align in the direction of the field but will randomize again at room temperature once the field is removed. These materials (such as molecular oxygen, gaseous nitric oxide, and aluminum) are termed "paramagnetic." And there is a third class of materials in which,

---

* Try this experiment: Take two credit card–size magnets, and hold them so that their magnetic sides face each other. When they are oriented so that both long sides point in the same direction, you can slide them smoothly past each other. Now flip one magnet, so that its long side is at a right angle to the long side of the other credit-card magnet. You should notice a bumpy, stick-slip motion as you now slide the magnets past each other. What you are feeling is the magnetic field of one card bumping over the walls between the stripes of magnetic domains on the other card.

due to the nature of the interactions between adjacent atoms and the chemical ordering of the atoms, their atomic magnetics (generated by electron orbits within the atoms) line up *opposite* to an external magnetic field. If the external magnetic field's north pole points up, the atomic magnet's north pole rotates to point down. These materials are called "diamagnetic," and they try to cancel out any external magnetic field. Water molecules are diamagnetic, and since we are primarily composed of water, so are we.

It is through our diamagnetism that Magneto is able to levitate himself and other people as shown in fig. 26. In moderate-strength magnetic fields, the atoms in your body are not susceptible to being polarized. The diamagnetic interaction is weak, such that at room temperature the normal vibrations of the atoms overwhelm the attempt to magnetically align them. In a very strong field, roughly 200,000 times greater than the Earth's magnetic field (and over 100 times larger than the field of a refrigerator magnet), the diamagnetic atoms in your body can be induced to all point in the same direction—opposite to the direction of the applied field. Just as two magnets repel if they are brought together so that their north poles are facing each other, the now magnetically polarized person will be repelled by the external magnetic field Magneto is creating—the very field that magnetically aligned the atoms in the first place. As Magneto increases the magnetic field he generates, the magnetic repulsion can become strong enough to counteract the downward pull of gravity. (That is, the upward force of the magnetic repulsion can be equal to or larger than the downward force of the person's weight, and there is then a net upward force on the person, lifting him off the ground.) It takes a very big magnetic field in order to accomplish this, and the heavier the person, the greater the effort. But it can be done, and the High Field Magnetic Laboratory at the University of Nijmegen in the Netherlands has amusing images and movies on their Web site of floating frogs, grasshoppers, tomatoes, and strawberries, demonstrating the reality of diamagnetic levitation.

* * *

If an electrical current generates a magnetic field, then could a moving magnetic field induce an electrical current in a nearby wire? The answer, as anyone who has read *X-Men* comics would know, is yes. In previous battles with the X-Men, and occasionally with puny humans, Magneto employed his mutant talent to transform any metal object into either an offensive weapon or a defensive shield. Magneto's power is most effective on metals that are already magnetic. There are only three elements (iron, cobalt, and nickel) that are magnetic at room temperature. Magneto can manipulate a steel girder into any form he wishes through the iron it contains, but his power would be limited on a gold wedding band, unless he is willing to expend an extreme effort in polarizing the normally diamagnetic material. But Magneto's real power lies not with his ability to exert forces on other magnetic materials, but his control over electrical currents.

For example, the mutant master of magnetism once constructed a computerized control panel that automated the power-dampening fields that keep the X-Men from interfering with his plan of mutant conquest of humankind. To prevent the X-Men from deactivating the device, Magneto configured it so that it has no buttons or knobs that can be reprogrammed. Magneto controls the panel by altering the electrical currents that flow through the circuits, affecting them through the magnetic fields that they create. Furthermore, by varying the magnetic field over the control panel, Magneto can cause these currents to flow.

How would a varying magnetic field create an electrical current? The answer brings us back to the point that started our discussion of electrical currents and magnetic fields: relative motion.

Just as a magnet can be pushed or pulled when a second magnet is brought near it, an external magnetic field can exert a force on an electrical current. As described in the previous chapter, moving electric charges generate a magnetic field that can interact with other magnetic fields, whether created by another electrical current or by a refrigerator magnet. When the charges are not moving, but are sitting still in a wire placed in an external magnetic field, there will be no force exerted on them.* What about if the external bar magnet is moved, while the charges remain sitting still in

---

* Careful readers will note that even if a wire is stationary (with respect to some observers), the electrons are in constant motion, zipping randomly to and fro, since they have a kinetic energy characterized by their temperature.

the wire? Assume that the magnet is moved toward the wire. From the magnet's point of view, it is not moving at all, but it is the wire that is moving toward it.

Magnetism is, at its heart, all about relative motion. If you were a blindfolded passenger in a car moving at constant speed in a straight line, how could you prove that when you arrived at your destination, it was the car that moved and not the scenery? If you change your speed or direction, then you will feel a force associated with the acceleration, and this will tip you off that it is you doing the moving. But for uniform motion you cannot really prove whether you or everything else are moving. All you can say for sure is that you are moving relative to your surroundings.

Similarly, when moving a magnet toward a wire, from the magnet's point of view it is stationary and it's the charges (both the mobile electrons and the fixed positively charged ions) in the wire that are moving toward it. But moving electrical charges create a magnetic field that will interact with the field of the magnet. So, by moving a magnet near a wire, the magnet sees two electrical currents from the positive ions and negatively charged electrons. A force is exerted on the charges in the wire and the electrons are free to move in response to this force. In this way Magneto is able to affect the direction of electrical currents in any device at will, though the precision by which he can guide them depends on how sensitively he can manipulate these magnetic fields.

If relative motion is the only factor that matters when considering whether a magnetic field affects electric charges, then how about a situation in which the magnet is stationary, but the wire is moved past the magnet? Would that generate a force on the charges?

To this physics answers, "Why not?" If I pull a wire through space, the electrons in the wire are moving, just as surely as if I kept the wire still and applied a voltage across it. In either case the electrons are moving past a fixed point at a certain speed. Relative to the magnet, it is as if there were an electrical current flowing by it, and we know how electrical currents and magnets interact. In

---

What is important for the discussion above is that in the absence of an external voltage or a changing magnetic field, there is no *net* motion of the electrons in the wire, so they can be considered, on average, to be stationary.

this situation a force will be applied to the charges in the moving wire that will induce them to flow. By dragging the wire through the external magnetic field, we convert the physical energy involved in moving the wire into a form of electrical energy manifested by the electrical current. For a coil of wire, it does not matter whether the magnet is pulled through the loop or the loop is moved past the magnet. As long as there is a relative motion between the charges in the wire and the magnitude of the magnetic field threading the loop, then a current will be induced, even without an outside voltage. This mechanism may sound a bit far-fetched, but it is in fact how the electricity coming into your house is generated.

A power station, such as the one Electro employs to charge up for a night of crime, operates on the principle that when a magnetic field passing through the plane of a coil of wire is changed, a current is induced in the wire. This is called Faraday's law, after Michael Faraday. The direction of this induced current is such that it creates a magnetic field that opposes the changing external magnetic field. This is a consequence of energy conservation, as we'll explain in a moment. In certain circumstances this current is termed an "eddy current," but it occurs whenever the magnetic field passing through a coil is increased or decreased.

Imagine a large magnet bent in the form of a broken ring, so the north pole faces the south pole, with a coil held in the open gap between its north and south poles. Initially the plane of the coil is at right angles to the magnet's poles, so the magnetic field threads through the loop. If the coil is now rotated 90 degrees, the plane of the coil faces away from the poles, so the amount of magnetic field passing through the coil is very small. Another right angle turn, and now the coil again faces the poles, and the magnetic field passing through it is large again. A further quarter rotation, and the field through the coil is minimum again, and so on. For every change in the magnetic field passing through the coil, whether an increase or a decrease, a current is induced. The direction of the induced current flips back and forth as the coil rotates. There are tricks for converting an alternating current (referred to as AC) to a direct current (known as DC). There are many practical reasons, that we won't go into now, for using AC to supply our electrical needs. The coils are made to rotate 60 times in one second, which is why in the United States the AC power has a frequency of 60 Hz

(Hz is an abbreviation for the unit of frequency "Hertz" and measures the number of cycles or rotations per second); while in Europe
the AC current's frequency is 50 Hz.

A current flows when the magnetic field passing through the
rotating coil changes. From a conservation of energy standpoint,
we realize that it must take energy to rotate the coil, in order to
create an electrical current that did not exist before. In *The Dark
Knight Strikes Again* # 1—Frank Miller's dystopian view of the future of the DC universe in which superheroes are forced into servi-
tude and Lex Luthor runs the country—the electricity that powers
a third of a large city was supplied by coercing the Flash to continually run on a treadmill. Recall from chapter 11 that the Flash has
managed to find a loophole around the principle of conservation of
energy (through his ability to tap into a "speed force"), so in Luthor's
view one might as well get some economic benefit from this suspension of the rules of physics. In our world, where we have yet to
find a single exception to the conservation of energy principle, the
energy that turns turbines and generates electricity comes from
the same process we use for making tea.

All commercial power plants generate electricity by boiling
water. The resulting steam turns a turbine (a fancy term for a pinwheel) to which the coils of wire in the powerful magnets are connected. As the turbines rotate, so do the coils, and electrical current
is produced. To boil the water, one either burns coal, oil, natural
gas, or garbage (termed bio-mass, since it sounds better). Alternatively, the excess heat generated from a nuclear reaction can boil
water and turn the turbine. But it all just goes toward creating
steam in order to turn a pinwheel attached to a coil between the
poles of a magnet. The stored chemical energy in the coal, oil, or
garbage has the same source as the chemical energy in the food we
eat—that is, from plant photosynthesis. The light from the sun is
a by-product of the nuclear fusion reaction running in the solar
core. (So, all electrical power plants could be viewed as nuclear
plants or solar plants, depending on your political bent.)

The turning of windmills results from temperature differences
in the atmosphere, arising from spatial variations in the sunlight
absorbed by the atmosphere or reflected by cloud cover. Obviously, solar cells (to be discussed in section three) require sunlight
in order to function. Aside from hydropower, in which the potential energy of water in a dam or waterfall is converted into kinetic

energy to turn a turbine, and geothermal power, in which the internal heat of the Earth is used to boil water, all other mechanisms for generating electricity involve the conversion of energy from the sun in one form or another. Clearly, without sunlight, none of us would be here. Perhaps the writers of *Superman* were on to something when they changed the source of Kal-El's powers from Krypton's excessive gravity to the light from our sun.

19

ELECTRO AND MAGNETO DO THE WAVE—

ELECTROMAGNETISM AND LIGHT

IN THE MID-1800S the expanding American frontier may not have seen many costumed crime-fighters, but there
was no shortage of heroes willing to fight for truth, justice, and the
Western Way. A good thing, too, as western comics would tap into
an upsurge in cowboys' popularity in the 1950s and help keep
comic-book publishers solvent during the superhero crunch pre-
cipitated by the campaign launched by Dr. Wertham's *Seduction
of the Innocent. All-American Comics,* featuring the adventures
of Green Lantern and the Justice Society of America, became *All-
American Western* starring the "fighting plainsman" Johnny Thunder (schoolteacher by day, gunfighter by night) and *All-Star Comics*
became *All-Star Western* with the Trigger Twins. In the DC universe, the scarred (physically and psychologically) rebel loner Jonah
Hex traveled across the western United States, righting wrongs and
saving widows. Similarly, Bat Lash and the Vigilante dispatched
justice . . . well, vigilante-style. Over in the Marvel universe, western comics were strictly kid stuff, with the Two-Gun Kid, Kid Colt,
Ringo Kid, and the Rawhide Kid working essentially the same beat,
moving from town to town every issue (though rarely ever bumping into one another), facing down rustlers and stagecoach robbers.
In the mid-nineteenth century as cowboy lawmen were cleaning
up the Wild West, both in our real-world universe and in western
comic books, physicists were elucidating the properties of electricity and magnetism, laying the foundation for our modern wireless lifestyle.

It was the Scottish physicist James Clerk Maxwell who, in

1862, as the Civil War raged on, made the monumental theoretical leap connecting electricity and magnetism and ushered in a new era of scientific advancement. The set of equations elucidated by Coulomb, Gauss, Ampère, and Faraday is nowadays known by the general title as "Maxwell's equations," for he recognized how they combined to provide a fundamental understanding of electromagnetic radiation. None of these scientists have ever starred in his own comic books, but without these heroes we'd still be reading by candlelight.

In order to understand how a toaster or electric lightbulb works, remember the water analogy we invoked earlier to explain electric currents: The water pressure of the faucet was the analog of an electrical voltage, while the amount of water per unit time flowing through a hose represented the electrical current. To indicate that the hose was not perfect, and that a finite pressure had to be continuously applied to maintain a steady flow through the hose, we suggested that the hose had both partially blocked regions as well as small pinholes along its length, through which water could escape and avoid participating in the main current flow. For a given hose's resistance, the greater the water pressure, the larger the water current. Alternatively, for a fixed water pressure, the larger the resistance, the smaller the current. These commonsense principles can be combined into a simple equation

$$\text{VOLTAGE} = (\text{CURRENT}) \times (\text{RESISTANCE})$$

known as Ohm's law, after Georg Ohm, another pioneering scientist in the early days of electromagnetism, for whom the basic unit of resistance is named. The longer and skinnier the hose, and the more clogs and holes along its length, the larger its resistance to current flow. A large water pressure at one end of a long narrow hose will correspond to a bare trickle at the other end, several miles away from the faucet. This is why your jumper cables are relatively short and thick, so that the current supplied from one battery is not degraded by the time it reaches the second battery.

The pinholes in the hose represent energy loss, and account, in our water analogy, for why a uniform water pressure (force) leads to a constant water current, and not an accelerating flow as Newton's second law would indicate. Copper wires obviously do not have holes through which the electrons leak out, but they do have resistance. At one end of the wire the electrons feel a large force due the accelerating voltage. Hence they have a large potential energy. As they flow down the wire, their potential energy is converted to kinetic energy. The greater their kinetic energy, the faster the electrons will move down the wire, and the larger the current. Imperfections or impurities in the wire act as speed bumps, and the fast-moving electrons collide with these defects, transferring some energy to them and causing the atoms to vibrate (which is why the wires become hot). For a given voltage, the resulting current is determined by the balance between the kinetic energy gained by the electrons from the applied voltage and the energy transferred to the imperfections.

These imperfections are impurities or atoms that are out of crystalline alignment with the rest of the lattice. As such, they have their own cloud of electrons surrounding them, just as the other atoms in the wire do. When these imperfections are shaking back and forth after collisions with the electrical current, their electric charges oscillate. Consider a variation on the swinging pendulum in chapter 9, in which the mass attached to a thin string now also carries an electric charge. As the charged mass swings back and forth, it is an electrical current, but one for which the speed of the charge's motion continuously changes. As such, it generates a magnetic field that is also changing in magnitude. However, a changing magnetic field induces an electric field. The oscillating charge will therefore continuously generate a varying electric field in phase with a changing magnetic field, radiating out into space. The faster the charged mass swings back and forth, the greater the frequency of the electric and magnetic oscillations created. Since the electric and magnetic waves have energy, the oscillating charged pendulum will slow down, even if there is no air resistance. There's a special name for the oscillating electric and magnetic fields created by the harmonic motion of the charged pendulum: It's called light.

WHY THE X-RAY SPECS ADVERTISED IN COMIC BOOKS ARE A TOTAL RIP-OFF

The oscillating impurity atoms in a wire carrying a current give off alternating electric and magnetic fields. The faster the electrons connected to the imperfections in the wire oscillate, the higher the frequency of the electromagnetic waves that are generated. Just by being at room temperature, all of the atoms (and their electrons) in the wire are vibrating at a rate of approximately a trillion cycles per second. Therefore any object at room temperature emits electromagnetic waves with a frequency of a trillion cycles per second. Electromagnetic waves with this vibration frequency are called infrared radiation. The greater the temperature, the faster the atoms in the object shake, and the higher the frequency of the emitted radiation.

Depending on how fast the charges in the atom are shaking—that is, how many times a second they move back and forth—the waves can have a wavelength (a measure of the distance peak to peak of the wave) of several feet to the diameter of an atomic nucleus. In the first case, we call these long-wavelength electromagnetic waves "radio waves" (with a frequency of roughly a million cycles per second) and in the second case, these ultrashort waves are termed "gamma rays" (with a frequency of over a million trillion cycles per second). Gamma rays have more energy and can therefore do more damage to a person than radio waves—as reflected in the fact that no one has ever gained superpowers by standing too close to an FM broadcast antenna. But in essence they are both the same phenomenon. In order for the atoms in a wire, such as the thin filament inside a lightbulb, to emit electromagnetic waves that our eyes are able to detect—which we call visible light—the atoms must be shaking back and forth roughly a thousand trillion times per second.

We are finally in a position to understand why the sun shines. As mentioned in chapter 2, the intense gravitational pressure at the center of the sun is great enough to force single protons (hydrogen nuclei) and neutrons together to form helium nuclei. The mass of a helium nucleus is slightly less than the two protons and two neutrons separately, and the mass deficit corresponds to a great energy output, through Einstein's expression $E = mc^2$. This outgoing energy balances the inward gravitational attraction, and the sun remains relatively stable as it burns through its fuel (and a lot of fuel it burns—600 million tons of hydrogen nuclei every second!). Part of the energy resulting from this fusion reaction is in the form of kinetic energy, and the rapidly moving charged helium nuclei emit electromagnetic radiation as they accelerate.

Acceleration is the rate of change in velocity, and so every time the helium nucleus either speeds up, slows down, or just changes direction as it collides with other nuclei in the dense stellar core, it emits light. It turns out that the light we see from the sun is very old, as it slowly makes its way from the center of the sun to the surface. It is difficult to see anything on a foggy night because the dense water-saturated atmosphere scatters the light in all directions. It's much denser within our sun, and the light generated from a nuclear fusion reaction takes an average 40,000 years before it is able to diffuse from the core to the surface of the sun.

Our eyes can see visible light because most of the light from the sun that makes it through the atmosphere is in this portion of the electromagnetic spectrum. When creatures without eyes evolved into creatures with eyes on this planet, the eyes they developed were most sensitive to the type of electromagnetic waves that were most prevalent. There are much fewer X-rays emitted by the sun compared to light in the "visible" portion of the spectrum. Consequently, if our eyes were tuned only for X-rays, we would live in a world of near-total darkness. Those creatures that *do* live in total darkness, such as sea life at the ocean's deep bottom where sunlight does not penetrate, do not waste genetic resources on superfluous eyesight or skin pigmentation, but rather rely on other senses to navigate.

Back in the Silver Age of comic books unscrupulous salesmen, preying on the prurient interests of comic-book readers, sold "X-ray glasses" that promised the wearer would be able to see through solid objects such as clothing. Even if these X-ray specs employed a principle similar to "night-vision" goggles, which translate infrared radiation into the visible (more on this in section three), there are not enough X-rays outside of a dentist's office to make this a useful product. And just try getting your money back from these companies!

Animals that are primarily nocturnal devote more of their optical receptors to high sensitivity rods, sacrificing color vision by having fewer cones, in order to detect the few electromagnetic waves present. But any animal or person that evolved "X-ray vision" would spend most of its time bumping into things, and would be at a distinct evolutionary disadvantage. The more electrons an atom has, the more strongly it scatters X-rays. This is why X-rays can penetrate through the soft tissue (which is mostly water) until reflected from the much denser bone. Presumably Superman is able to emit X-rays from his eyes, which then penetrate through low X-ray absorbing matter before being reflected and then detected by the Man of Steel. Those not from Krypton can only see when light from an external source is reflected into their eyes from an object. It costs the human body energy and raw materials to develop optical nerve cells sensitive to low wavelength light; consequently there is little point in developing an ability to detect the odd occasional X-ray.

I CAN TELL THAT YOU'RE THINKING YOU WISH YOU HAD A TIN-FOIL HELMET

The leader of the mutant team of superheroes known as the X-Men is the wheelchair-bound telepath Charles Xavier, also known as Professor X. While his shattered spine may have left him unable to walk, he was a formidable general for this team of "good" mutants thanks to his ability to read and project his thoughts into other people's minds. The physical basis underlying Prof. X's telepathy (and that of his protégée Jean Grey, and Saturn Girl of the Legion of Super-Heroes, for that matter) is that time-varying electrical currents can create electromagnetic waves detectable to someone who is supersensitive.

All cells in our body have a function. Muscle cells exist to generate a force, whether it is for the flexing of the biceps or the pumping of the heart. Liver cells filter impurities from the bloodstream, while stomach and intestinal cells put them there in the first place. The role of nerve cells or neurons is to process information. One way they accomplish this is by transmitting and altering electrical currents. The charged objects that move from neuron to neuron are not electrons, but calcium, sodium, or potassium atoms that are either missing one or more of their electrons or have acquired extra electrons (such charged atoms are termed "ions"). An accumulation of ions in one region in the brain creates an electric field that in turn coerces other ions in other neurons to move. The moving ions constitute a current that generates a magnetic field. Experiments by neuroscientists using sensitive electrodes placed within the brain can detect the electric fields generated by the motion of these ions, which typically vary randomly in time.

Depending on where in the brain the electrode is located and what task the brain is performing, the electric fields recorded will assume a coherent wave form, oscillating through several periodic cycles before abruptly returning to the random background. Neuroscientists are beginning the difficult task of identifying the voltage variations and determining their significance (if any) with behavioral tasks. From such simple building blocks the human mind, with all its vast complexity, is constructed.

While scientists are a long, long way from understanding how or whether the ionic currents in the brain lead to consciousness, there is one aspect of the neuronal currents that we may rely upon: that moving electric charges generate magnetic fields. In turn, because the ionic currents in the brain are continually changing direction and magnitude, the corresponding magnetic fields vary in time and create electric fields as well. The net effect is that very low-frequency electromagnetic waves radiate out from the brain whenever electrical activity occurs. The wavelengths, amplitudes, and phases of these electromagnetic waves are determined by the time-dependent ionic currents from which they originate. The amplitude of these waves is extremely weak, such that their power is over a billion times lower than the background radio waves that surround you at this very moment (ordinarily, the fact that we live within a sea of radio broadcast signals is ignored, until one turns on a radio and can't tune in a particular station clearly). Yet the electromagnetic waves created by brain currents do exist, though their intensity is too weak to be observed unless the sensor is directly on the person's head. For certain powerful mutants such as Prof. X, or residents of the moon Titan in the thirtieth century (Saturn Girl), their miracle-exception involves brains sensitive enough to detect the electromagnetic waves generated by others' thoughts. Of course if you are wearing a metal helmet (a precaution taken by Xavier's evil stepbrother, the Juggernaut; Magneto; and other farsighted X-Men foes) then your head is shielded so that the outgoing (and any incoming) electromagnetic waves are grounded out.

A stone tossed into a pond creates a series of ripples that grow weaker the farther one is from the source. The water molecules have a great deal of kinetic energy, imparted from the falling stone. But as the ripple becomes larger and larger, the amount of energy in the water molecules is spread out along the growing circumference. The kinetic energy of the water molecules per unit length on the ring's border is diluted as the wave propagates outward, so that for a stone dropped in the middle of the Pacific Ocean there is no noticeable disturbance at the California coast. In the same manner there is a decrease in electromagnetic waves' intensity the farther one moves from their source. The fact that the intensity of electromagnetic waves decreases with distance from the source of the waves explains why, when Prof. X needs to locate a particularly distant mutant, he uses an electronic amplifier of his mental powers, called Cerebro. First introduced in *X-Men* # 7 as an automated mutant brain-wave detector, it was adapted in later issues to increase the sensitivity of Prof. X's telepathic powers. The recognition that in order to detect a distant electromagnetic signal, one would have to use an external amplifier is consistent with the physical mechanism underlying Prof. X's mutant power. This is also why radio and television broadcast stations use megaWatts of power to transmit their signals. A Watt is a unit of power, defined as energy (in Joules or kg-m^2/sec^2) per second, and a megaWatt is a million Watts. The more power a radio station can broadcast, the greater will be the intensity of the electromagnetic waves reaching a given remote antenna, and the stronger will be the signal received by the radio. Commercial radio stations generate their signals by oscillating charges up and down a big antenna. Your television set or radio does not employ Cerebro technology to amplify the distant signal, but it does make use of transistors for this function, which we will discuss in detail in chapter 23.

* * *

Detecting electromagnetic waves created by someone's thoughts is one thing, but can one work backward to determine the neuronal currents that generated them—that is, can one actually read and interpret someone else's thoughts? Yes. Prof. X and Saturn Girl presumably do it the same way that "reverse-television" functions. Let me explain how this would work.

Television signals consist of electromagnetic waves sent out by a powerful transmitter that, upon striking a rooftop antenna, cause the charges to oscillate back and forth with a frequency and amplitude characteristic of the incident signal. The information encoded within the electromagnetic wave is then sent to the television set. The heart of the set is the picture tube consisting of a large glass surface, onto which is evaporated a phosphorescent material that gives a brief flash of light when struck by an energetic electron. This glass face is one end of an irregularly shaped glass box. At the narrow end of the box is a wire, heated by an external voltage so that electrons are stripped from it. These now-free electrons are then directed, by way of metal "steering" plates at suitable voltages, toward the other end of the picture tube—that is, the end having the large face coated with phosphorescent material. By choosing the right voltage on the steering metal plates, the electrons can be directed to strike a particular region of the screen. The interior of the television tube is evacuated, so as to reduce the number of stray air molecules that could cause unwanted scattering of the electron beam. Whenever the beam strikes the screen, depositing its kinetic energy into the phosphor material, it causes a flash of light to be emitted. The voltages applied to the steering plates are then adjusted, and the electron beam is now directed to another location on the screen, lighting another phosphor or leaving it dark if the beam is stopped. This continues until the electron beam has moved across the entire screen. A given array of lit and dark regions across the screen yields an image on the face of the television.

By slightly changing the image being projected onto the screen, the illusion of motion can be induced. A radio broadcast simultaneous with the transmission of the visual signal provides the sound. If three different phosphors or colored filters emitting red, green, and blue light are used at the spot struck by three electron beams, then slight adjustments as to how much of each filter is illuminated in each location results in a color picture. The basic physics underlying television is that the information encoded in the electromagnetic wave contains a set of instructions for the magnitude and timing of the voltages to be applied to the steering plates.

The varying electron beam in the picture tube gives off its own set of electromagnetic waves, different from the waves that were received by the antenna, but related to the television image. A sensitive antenna placed near this monitor could detect these electromagnetic waves and, with the appropriate software, reconstruct the image that the electron current is intended to create. This "reverse-television" is a very inefficient way to have two sets showing the same image, but it would be one method by which a person could read the information projected onto a computer monitor without directly hacking into the computer. Or by which to send information from one brain to another.

What about Prof. X using the power of his mutant mind to control the actions of someone else? Recent experiments suggest that this situation may not be so far-fetched. It has been demonstrated that not only can we detect the weak magnetic fields created by ionic currents in the brain, but also that the reverse process is possible as well. Neuroscientists have developed a research tool called Transcranial Magnetic Stimulation (TMS). In this procedure a randomly varying magnetic field is applied to a human test subject's head, providing electrical stimulation to selected regions of the cerebral cortex. The subject's reaction time and ability to initiate a voluntary hand movement are disrupted by the application of the external magnetic field.

The ability to control the actions of others, using just the power of one's mind, is not limited to mutants and thirtieth-century heroes from Saturn's moon Titan. In fact, I have been known to demonstrate just such amazing mental powers. Often my lectures can induce my students to either exit the lecture hall at a great speed or to fall into a deep and profound slumber!

SECTION
3

MODERN PHYSICS

20

JOURNEY INTO THE MICROVERSE—
ATOMIC PHYSICS

COMIC-BOOK READERS have a reasonable expectation that the hero will be triumphant by the end of the story. The fun, therefore, lies in the challenges that must be overcome as each month's tale races toward its conclusion. Not to put too fine a point on it, but the better the supervillain, the better the story—which is probably one reason the *Fantastic Four* was such a popular comic book in the early sixties. Certainly the artwork by Jack Kirby was a major factor, as was Stan Lee's plotting and characterization of the intrepid quartet. But if a superhero is only as good as his or her nemesis, then the Fantastic Four achieved greatness in issue # 5 (July 1962) when they became "Prisoners of Doctor Doom."

Victor von Doom was a scientific genius second only to Reed Richards, the leader of the Fantastic Four. Richards and von Doom attended the same college, both on "science scholarships" (in the fantasy world of comic books, institutions of higher learning compete for academic scholars the way our real-world universities vie for athletes). Von Doom was expelled when one of his "forbidden" scientific experiments went disastrously awry, blowing up the lab and scarring his face. Hiding his disfigurement behind a metal faceplate, he designed a high-tech suit of armor that rivaled Iron Man's and began a long campaign of world conquest as Doctor Doom. Of course, not having finished his degree, von

Doom is not really a doctor, and it is most likely his bitterness about his A.B.D.* status, along with his desire to humiliate Reed Richards, that drives his evil ambitions. Unlike the villains in DC stories in the 1960s, who would inevitably be captured and turned over to the police at story's end, the Fantastic Four never seemed to be able to battle Doctor Doom to better than a draw. Of course, because Doom was the dictatorial ruler of the small European nation of Latveria, it was always a bit vague as to who the proper authorities were that one could hand a head-of-state over to.

More to the point, Doom's pride was so great that he would rather face near-certain death than incarceration. Consequently, a typical battle with Doctor Doom would end with Doom either being lost in space, marooned in another dimension, or trapped in time—all fates he had intended for the Fantastic Four. At the climax of the aptly titled "The Return of Doctor Doom" in *Fantastic Four* # 10, Doom was struck by a reducing ray he had planned to use on the Fantastic Four. The story ended with Doom shrinking to nothingness, but this would not be the last we would see of Doom. Six issues later in *Fantastic Four* # 16 the Fantastic Four journeyed to "The Micro-World of Doctor Doom," where they learned that Doom had survived his shrinking ordeal. At some point in his reduction, Doom entered a "micro-world—a world that might fit on the head of a pin." Later, in *Fantastic Four* # 76, Reed, Ben, and Johnny ventured into an entire microverse, that is, a universe (or at least a galaxy) of micro-worlds. The microverse was depicted as residing within a stain on a microscope slide in Reed Richards's laboratory. This at least removed the need to explain the enormous coincidence in *Fantastic Four* # 16 of Doctor Doom and the FF standing exactly above one such micro-planet just as they began to shrink.

If the micro-world that Doom encountered and subsequently conquered could indeed fit on the head of a pin, then its diameter at the equator is approximately 1 millimeter. For comparison, the Earth's diameter is 13,000 kilometers. One kilometer is one million millimeters long, so the micro-world is thirteen billion times smaller than the Earth. Recall that in chapter 7 we discussed the

---

* All But Dissertation.

difficulties involved in reducing an object in size. The micro-world cannot be six billion times denser than our planet, unless it is composed of white-dwarf-star matter. The fact that Doctor Doom, the Fantastic Four, and the inhabitants of this micro-world are able to walk around normally suggests that this is not the case. Both Doctor Doom and Reed Richards seem as smart on the micro-world as they do at their normal size, and the Thing is not any less strong, so it is unlikely that atoms are removed from them upon shrinking. We must therefore regretfully conclude that the micro-world of Doctor Doom is much like his other "master plans"—impressive in principle but disappointing in execution.

* * *

If it's that hard to construct a world to fit on the head of a pin, what are we to make of the Atom's adventure with "The Deadly Diamonds of Doom" in *Atom* # 5? In this story a diamond artifact found on Mount Pico in the Azure Islands is brought back to Ivy Town by an archeologist friend of Ray Palmer (alter-ego of the Atom), whereupon a strange ray beam emanates from it, turning people and house cats into diamond statues. *Even though it appears solid*, Professor Palmer thinks as he reaches for his size and weight controls, *there are vast gulfs of space between the atoms that comprise the diamond!*

True enough. Most of an atom is indeed empty space between the positively charged nucleus and the average location of the negatively charged electrons. As the Tiny Titan shrinks down to subatomic length scales, he discovers an entire other planet, *inside an atom*! I'm at a loss to suggest of what this other planet could be composed. It certainly can't be made of atoms, since it is smaller than the electrons that reside in the diamond artifact. Certainly the discovery that there could be entire civilizations residing within the atoms of ordinary matter would garner Palmer a Nobel Prize at the very least, along with worldwide fame and fortune. Such is the hallmark of this hero that he never even considers reporting this scientific discovery, nor any of the other micro-worlds he encounters in *Atom* comics # 4, # 19, the *Justice League of America* # 18, or *Brave and the Bold* # 53.

While comic-book claims that there are micro-worlds within

atoms is pure fantasy, the region within an atom as elucidated by quantum mechanics is no less strange. Within the "empty spaces" inside an atom are the "matter-waves" associated with an electron's motion. These matter-waves are the key to understanding atomic physics.

WHAT DO YOU DO WHEN EVERYTHING YOU KNOW IS WRONG?

It is now time for us to delve into the world of atoms. Things will get physics-y here for a few pages, but bear with me. We'll get back to comic books soon enough, but some background is needed in order to understand why at least some physicists take the notion of parallel universes and an infinite number of Earths seriously.

At the end of the nineteenth century, there was a growing body of experimental evidence that indicated that the physical principles described in the preceding chapters failed at explaining the behavior of atoms and light. For example, physicists were stymied trying to explain why hot things glowed. Place an iron poker in a roaring fireplace and, as it warms, it initially glows red and eventually gives off white light. Thanks to Maxwell's theory of electromagnetism discussed in the previous chapter, physicists understood that the oscillating electric charges in each atom, shaking back and forth as the poker became hotter, emitting light, and that the faster the atoms shook back and forth, the higher the frequency would be of the resulting electromagnetic radiation. Back in the 1800s scientists had devised very clever techniques for measuring both ultraviolet and infrared light, at the upper and lower regions of the visible electromagnetic spectrum, bracketing the narrow slice of light that our eyes can see. Consequently, they could accurately measure exactly how much light a hot object emitted at any given wavelength as its temperature was increased. They discovered two surprising things. First, the amount of light emitted at a specific wavelength depends only on the temperature of the object, and not on any other characteristic. Regardless of an object's material composition, shape, or size, the only thing that determined the spectrum of light emitted was its temperature.

Secondly, the total amount of light emitted was not infinite and also depended only on the temperature. This second point was the first falling domino that ultimately led to the development of quantum mechanics.

The fact that the light from a hot object depends only on its temperature is a consequence of the principle of conservation of energy. If two objects made of different materials at the same temperature emitted different radiation spectra, there would be a way to have a net transfer of energy between them and hence useful work, without any heat flow. While this would be a convenient violation of the second law of thermodynamics, it does not occur for just that reason. A practical benefit of the fact that the emitted light spectrum depends *only* on the temperature is that we can use the intensity of emitted light as a function of wavelength to determine the temperature of objects for which normal thermometers are useless. This is how the surface temperature of the sun (roughly 11,000 degrees Fahrenheit) and the background microwave radiation remnants of the Big Bang (3 degrees above Absolute Zero) are measured through observations of the spectrum of light they produce.

The second discovery, that the energy emitted by a glowing object is not infinite, was not really a shock to physicists. What they found disturbing was that Maxwell's electromagnetic theory predicted that the amount of light energy emitted should increase without limit. Calculations using Maxwell's theory correctly predicted how much light would be emitted at low frequencies, in exact agreement with observations. As the frequency of light emitted by a hot object increased into the ultraviolet portion of the spectrum, the measured light intensity reached a peak and at higher frequencies decreased again back to a low value, which is what one would expect from both conservation of energy and common sense. However, the calculated curve indicated that the intensity would become infinitely high with rising frequency. This was labeled the "ultraviolet catastrophe," though it was only a "catastrophe" for the theorists doing the calculations. Many scientists checked and rechecked the calculations, but they could find nothing wrong.

Maxwell's equations had worked so well in all other cases (they led to the invention of radio in 1895 and would eventually enable the development of television as well as all forms of wireless communication) that it was doubtful that there was something fatally wrong with them. Rather, scientists concluded that the problem must lie with applying Maxwell's theory to the shaking atoms in a glowing object. Again, many tried to find an alternative approach, some different theory that could account for the observed spectrum of light emitted by a glowing object. Here is where the fact that the spectrum depends only on the temperature of the object becomes important. If the theory of electromagnetism could not account for the behavior of one or two exotic pieces of matter, well, that would be somewhat awkward, but not earthshaking. This inability to explain a property shared by *all* matter was downright embarrassing, and something had to be done.

In 1900 the theoretical physicist Max Planck, recognizing that desperate times called for desperate measures, did the only thing he could to explain the spectrum of light emitted by a glowing body: He cheated. He first determined the mathematical expression that corresponded to the experimentally obtained glow curve. Once he knew what formula he needed, he then set out to find a physical justification for it. After trying various schemes, the only solution he could come up with that gave him the needed glow-curve equation involved placing restrictions on the energy of the atoms that made up the glowing object. Planck proposed that the electrons in any atom could only have specific energies. From the Latin word for "how much," this theory was called quantum physics. The separation between adjacent energy levels was in practice very small. And I mean *really* small: If the energy of a well-hit tennis ball was 50 kg-meter^2/second^2, then the separation between adjacent energy levels in an atom is less than a millionth trillionth of a kg-meter^2/second^2. This should provide some perspective the next time you hear a commercial boasting that the latest innovation in an automobile or laundry detergent represents a "quantum leap."

Planck had to introduce a new constant into his calculations, an adjustable parameter that he labeled "h." He assumed that any change in the energy of an atom could only take on values $E = hf$, or $E = 2hf$ or $E = 3hf$, and so on, but nothing in-between (so the atom could *never* have an energy change, say, of $E = 1.6hf$ or $17.9hf$) where f is the frequency characteristic to the specific atomic element. This is like saying that a pendulum can swing with a period of one second to complete a cycle or ten seconds, but that it was impossible to make the pendulum swing back and forth in five seconds. Planck himself thought this odd, but was necessary in order to make his calculations come out right. He intended to let the value of **h** become zero once he obtained the correct expression for the spectrum for a glowing object. To his dismay he discovered that when he did this, his mathematical expression went back to the infinite energy result from classical electromagnetism. The only way to avoid this nonsensical infinite result is to say that the atoms *cannot* take on any energy value they want, but must always make changes in discrete steps of magnitude $E = hf$. Since **h** is very, very small (**h** = 660 trillionth trillionth trillionth of a kg-meter^2/sec), we never notice this "graininess" of energy when we deal with large objects such as baseballs or moving automobiles. For the energy scale of an electron in an atom it is quite significant and absolutely cannot be ignored.

The fact that the energy of electrons in an atom can only have discrete values, with nothing in between, is indeed bizarre. Imagine the consequences of this discreteness of energy for a car driving down the highway at 50 mph if Planck's constant **h** were much larger. The quantum theory tells us that the car could drive at a slower speed of 40 mph, or at a faster speed of 60 mph, but not at any other speeds in between! Even though we can conceive of the car driving at 53 mph, and calculate what its kinetic energy would then be, it would be physically impossible for the car to drive at this speed, according to the principles of quantum physics. If the car absorbed some energy (say from a gust of wind), it could increase its speed to 60 mph, but only if the energy of the wind could exactly bridge the difference in kinetic energies. For a slightly less-energetic gust, the car would ignore the push of the wind and continue along at its original speed. Only if the energy of the wind exactly corresponded to the difference in kinetic energy from 50 to 60 mph, or 50 to 70 mph, would the car "accept" this push, and move to a higher speed. The transition to the higher velocity would be almost instantaneous, and the acceleration during this transition would do bad things to the car's occupants. This scenario seems ridiculous when translated to highway traffic, but it accurately describes the situation for electrons in an atom.

Is there any way to understand why the energy of an electron in an atom has only certain discrete values? Yes, actually, but first you must accept one very strange concept. In fact, all of the "weirdness" associated with quantum physics can be reduced to the following statement: *There is a wave associated with the motion of any matter, and the greater the momentum of the object, the shorter the wavelength of this wave.*

When something moves, it has momentum. The physicist Louis de Broglie suggested in 1924 that associated with this motion, there is some sort of "matter-wave" connected to the object, and the distance between adjacent peaks or troughs for this wave (its wavelength) depends on the momentum of the object. Physicists refer to an object's "wave function," but we'll stick with "matter-wave" as a reminder that we are referring to a wave associated with the motion of a physical object, whether an electron or a person.

This matter-wave is *not* a physical wave. Light is a wave of alternating electric and magnetic fields created by an accelerating electric charge. The wind-driven ripples on the surface of a pond or the concentric rings formed when a stone is tossed into the water result from mechanical oscillations of the water's surface. Sound waves are a series of alternating compressions and expansions of the density of air or some other medium. In contrast, the matter-wave associated with an object's momentum is not like any of these waves, but in some sense it is just along for the ride, moving along with the object. It is not an electric or magnetic field, nor can it exist distinct from the object or does it need a medium to propagate. Yet this matter-wave has real physical consequences. Matter-waves can interfere when two objects pass near each other, just as when two stones are thrown into a pond a small distance apart, each creates a series of concentric ripple rings on the water's surface that form a complex pattern where the two rings intersect. If you ask any physicist what this matter-wave actually *is*, he or she will give a variety of mathematical expressions that always boil down to the same three-word answer: *I don't know.* For once, our one-time "miracle exception" applies to the real world, rather than the four-color pages of comic books!

Unless an object is moving near the speed of light, its momentum can be described as the product of its mass and its velocity. A Mack truck has more momentum than a Mini-Cooper, if they are both traveling at the same speed, since the truck's mass is much larger. The Mini-Cooper could have a larger momentum if it were traveling at a much, much higher speed than the truck. Physicists typically use the letter "p" to represent an object's momentum, since *obviously* the p stands for *mo*-mentum.* The wavelength of this matter-wave is represented by the Greek letter lambda λ. The matter-wave's wavelength was proposed by de Broglie (and experimentally verified in 1926 by Clinton Davisson and Lester Germer) to be related to the object's momentum by the simple relationship (momentum) times (wavelength) equals a constant, or $p\lambda = h$ where h is the same constant that Planck had to introduce in order to account for the glow curve of hot objects.

The fact is that the product of an object's momentum and the matter-wave's wavelength is a constant means that the bigger the momentum, the smaller the wavelength of the matter-wave. Given that momentum is the product of mass and velocity, large objects such as baseballs or automobiles have very large momenta. A fastball thrown at one hundred mph has a momentum of about 6 kg-meter/second. From the relationship $p\lambda = h$, since h is so small, this indicates that the wavelength (the distance between successive peaks in the wave, for example) of the matter-wave of the baseball is less than a trillionth trillionth of the width of an atom. This explains why we have never seen a matter-wave at the ballpark. Obviously there is no way we can ever detect such a tiny wave, and baseballs, for the most part, are well-behaved objects that follow Newton's laws of classical physics.

On the other hand, an electron's mass is very small, so it will have a very small momentum. The smaller the momentum, the bigger the matter-wave wavelength, since their product is a constant. Inside an atom the matter-wave wavelength of an electron is about the same size as an atom, and there is no way one can ignore such matter-waves when considering the properties of atoms. When the DC Comics superhero the Atom shrinks down to the size of an atom, he should see some rather strange sights. At this size he is smaller than the wavelength of visible light so, just as we can't see radio waves, whose wavelength is in the range of several

---

* Seriously, you wouldn't want to represent momentum by the letter "m," for that would lead to confusion as to whether the "m" referred to momentum or mass. I'm not really sure why the letter "p" has been associated with momentum, but it seems to have stuck.

inches to feet, the Atom's normal vision should be inoperable, and he will be roughly the same size as the matter-waves of the electrons inside the atom. It is suggested in his comic that at this size the Atom's brain interprets what he sees as a conventional solar system image, for he has no valid frame of reference to otherwise decipher the signals sent by his senses.

Imagine an electron orbiting a nucleus, pulled inward by the electrostatic attraction between the positively charged protons in the nucleus and the electron's negative charge. As the electron travels around the nucleus, only certain wavelengths can fit into a complete cycle. When the electron has returned to its starting point, having completed one full orbit, the matter-wave must be at the same point in the cycle as when it left. As weird as the notion of a matter-wave is, it would be even harder to comprehend if when the wave left it was at a peak (for example), and after having completed one full orbit, was now at a valley. In order to avoid a discontinuous jump from a maximum to a minimum whenever the wave completed a cycle, only certain wavelengths that fit smoothly into a complete orbit are possible for the electron. This is not unlike the situation of a plucked violin string, with only certain possible frequencies of vibration. Because the wavelength of the matter-wave is related to the electron's momentum, this indicates that the possible momenta for the electron are restricted to only certain definite (discrete) values. The momentum is in turn related to the kinetic energy, so the requirement that the matter-wave not have any discontinuous jumps after finishing an orbit determines that the electron can only have certain discrete energy values within the atom.

These finite energies are a direct result of the constraint on the possible wavelengths of the matter-waves, which in turn are due to the fact that the electron is bound within the atom. An electron moving through empty space has no constraints on its momentum, and consequently its matter-wave can have any wavelength it likes.* A piece of string can have any shape at all when I wiggle one end, provided the other end is not attached to anything. But if the string is clamped at both ends, as in the case of a violin string,

---

* Therefore our earlier analogy with an automobile driving on a highway that can only have speeds in multiples of 10 mph would only be valid for a car doing laps on a closed track.

then the range of motions for the string are severely restricted. When I now pluck the clamped string, it can only vibrate at certain frequencies, determined by the length and width of the string and the tension with which it is clamped. There is a lowest fundamental frequency for the string and many higher overtones, but the string cannot vibrate at any arbitrary frequency once it is constrained in this manner.

Likewise the electron is held in an orbit by its electrostatic attraction to the positively charged nucleus. If "plucked" in the right way, a matter-wave for the bound electron can take on a higher energy value. When the electron then relaxes back to its lower fundamental frequency, it must do so by making a discrete jump. Energy is conserved; consequently, the electron can only lower its energy when returning to the lower frequency level by giving off a packet of energy equal to the difference between its higher energy level and the lower level it is relaxing to. Because the energies available to the electron are discrete, well-defined values similar to the overtones possible for a clamped string, this jumping from one energy state to another is termed a "quantum transition" or a "quantum jump." The discrete packet of energy given off by the electron when making this transition is typically in the form of light, and a quantum of light energy is termed a "photon" (a concept introduced by Albert Einstein, again in 1905—a busy year for him and physics—though the term "photon" was not coined until 1926 by Gilbert Lewis).

If a glass tube is filled with a gas such as neon, and an electrical current is passed through the gas, the energetic electrons of the current will sometimes collide with the neon atoms. When the energy of the energetic electrons is just right, the neon atoms can be excited into a higher energy state. After the collision, the excited neon atoms will relax back to their initial lower energy configuration, emitting a photon of light that has the frequency (hence *color*) that corresponds to the energy difference between its starting and final states. This is why neon lights have their identifiable color. By changing the type of gas in the tube, different colors of light can be selected. You could do this with any gas, but only certain elements have a transition within the visible portion of the light spectrum. If the atoms suffer highly energetic collisions, such that many higher energy states are excited, then many discrete wavelengths of light will be given off when the different overtones all relax back to the fundamental level. Different elements have differing arrays of overtones and fundamental frequencies, just as different strings on a violin or guitar will have different vibrational modes depending on the length, width, and tension. Two identical violin strings clamped with the same tension will have the same range of possible frequencies when plucked. Similarly, two identical atoms will have the same spectrum of emitted light when they relax from an excited state. In this way the spectrum of wavelengths of light emitted by an energetic atom is unique and can be considered the fingerprint of the particular element. The lighter-than-air element Helium was discovered by the detection of its characteristic spectrum of light coming from the sun (the word "helium" is derived from Helios, the Greek sun god). By careful comparison with the spectrum of light emitted by hydrogen and other gases, scientists concluded that this array of wavelengths must arise from a new element that at the time had never been found terrestrially. Fortunately for the Macy's Thanksgiving Day Parade, underground pockets of helium were eventually discovered.

The notion that there is a wave associated with the motion of any object, and that the wavelength of this wave is inversely proportional to its momentum, is weird, but by accepting this mysterious concept we gain an understanding for the basis of all of chemistry. Bring two atoms close enough together, and they may form a chemical bond and in so doing create a new basic unit, the molecule. Why would the atoms do this? The negatively charged electrons in the first atom certainly will repel the negatively charged electrons in the second atom. Before quantum mechanics, there was no satisfactory fundamental explanation for why the universe didn't just consist of isolated elemental atoms.

The driving force underlying the bonding between atoms is the interactions of the matter waves of the electrons from the different atoms. When the two atoms are held very far apart, the matter waves of the atomic electrons do not overlap. When the atoms are brought close enough together so that the electron clouds around each atom intersect, the electronic matter waves from each atom begin to interfere, forming a new wave pattern, just as two stones tossed into a pond create an intricate pattern of ripples that is very different from the pattern that would be created by each stone separately. In most cases this new pattern is a high-energy, discordant mess, similar to the sound resulting from a clarinet and violin played simultaneously by novices with no musical training or talent. In these cases the two atoms do not form a chemical bond and do not chemically interact. In a few special cases the two matter waves interact harmoniously, creating a new wave pattern that has a lower-energy configuration than the two separate matter waves. In these special cases the two atoms can lower their total energy by allowing the matter waves to interact in this manner, and once in a lower-energy state, it requires the addition of energy to physically separate them. In this way, despite the considerable repulsion between the negatively charged electrons, the two atoms are held together in a chemical bond, owing to the wavelike nature of the electrons.

These arguments about discrete energy levels in an atom arising from those particular orbits that correspond to an integral number of wavelengths of the electron matter-wave seem so reasonable that it is a shame that they're not really correct. The electron cannot literally be considered to move in a circular or elliptical orbit around the positively charged nucleus, despite the appealing analogy with our solar system. For one thing, the electron would be constantly accelerated as it bends onto a curved path. As argued in the previous chapter, an accelerating electric charge in a circular orbit emits electromagnetic waves that carry energy, so as the electron radiates light in its orbit it loses kinetic energy. Eventually the electron should spiral into the nucleus in a little less than a trillionth of a second, so no elements should be stable, and hence there would be no chemistry and no life if the electrons really moved in curved orbits.

Nevertheless, the notion of only certain wavelengths being allowable, with corresponding discrete energy levels, is still a valid one, even if the picture we employed to get there can only be considered a useful metaphor and not a literal description. Instead of thinking of the electron as a point particle that moves in a circular orbit with a particular matter-wave associated with it, the full quantum theory of Heisenberg and Schrödinger, to be discussed in the next chapter, tells us that there is a "wave function" for the electron. Just as, for the case of the plucked violin string, it makes no sense to ask where exactly on the string the wave is, similarly for the electron in an atom, its matter-wave extends over the atom, and we cannot specify the electron's position or trajectory more accurately than this. Electrons only emit or absorb light when moving from one wave pattern to another in the atom. As we'll see in the next chapter, these matter-waves are also responsible for a crisis on infinite Earths!

NOT A DREAM! NOT A HOAX! NOT AN IMAGINARY TALE!–

QUANTUM MECHANICS

THE ORIGIN TALE in *Showcase* # 4 describing how Barry Allen gained his super-speed powers and became the Silver Age Flash featured a symbolic passing of the torch from the earlier Golden Age of superheroes. Just prior to being struck by the lightning bolt that simultaneously doused him with exotic chemicals, police scientist Allen relaxed with a milk-and-pie break in his lab while reading *Flash Comics* # 13, featuring the Golden Age Flash on its cover. After the freak accident gave Barry his superpowers, his immediate thoughts turned to how he could use his super-speed to help humanity. Taking his inspiration from the Flash comic book he had been reading before he was struck by lightning, he donned a red-and-yellow costume and began his crimefighting career as the Silver Age Flash (though he referred to himself simply as the Flash, not realizing that he was himself a comic-book character in the dawning Silver Age of superheroes). In a turn that nowadays would be described as "postmodern," and back then was simply considered a "neat idea," it was established in the Flash comic books of the 1960s that the Flash character of the 1940s (who wore a different costume and gained his super-speed in a different, though equally implausible, chemical accident) was a comic-book character in Barry Allen's "reality."

The Golden Age Flash (whose secret identity was Jay Garrick) was considered fictional as far as the Silver Age Flash was concerned until September 1961. In the classic story "Flash of Two Worlds" in *The Flash* # 123 (fig. 27), it was revealed that the Silver Age Flash and the Golden Age Flash both existed, but on *parallel*

Fig. 27. *The cover to* The Flash # 123, *giving comic-book fans their first hint that there were at least two worlds beyond their own.*

Earths, separated by a "vibrational barrier." In this story the Silver Age (Barry Allen) Flash accidentally vibrated at super-speed at the exact frequency necessary to cross over to the Earth on which his idol the Golden Age (Jay Garrick) Flash lived. Once he realized that he was in the world of the Golden Age heroes, Barry met Jay and introduced himself. "As you know," the police scientist explained, "two objects can occupy the same space and time—if they vibrate at different speeds!" Apparently Barry Allen was a better forensic scientist than he was a theoretical physicist. Regardless of their vibrational frequency (and as we saw in section two, atoms in a solid do vibrate simply because they have some nonzero temperature), there is no way that two objects can be in the same place in space and time (unless we are discussing massless quantities such as light photons).

The writer of the "Flash of Two Worlds" story was Gardner Fox, who had also written many of the Golden Age Flash comics. He proposed a mechanism to explain how the Silver Age Flash could read comic books featuring the Golden Age Flash on this second Earth, and he also provided some insight into his work habits. As Barry hypothesized, "A writer named Gardner Fox wrote about your adventures—which he claimed came to him in dreams! Obviously when Fox was asleep, his mind was 'tuned in' on your vibratory Earth! That explains how he 'dreamed up' The Flash!"* This crossover meeting between the Silver Age and Golden Age Flash was a hit with comic-book fans, and the Silver Age Flash would more and more frequently cross the vibrational barrier to Earth-2. The world on which the Golden Age Flash resided, though it appeared first chronologically, was labeled Earth-2, while the world of the Silver Age was designated Earth-1. The reader's world, in which all superheroes exist solely as fictional characters in comic books, was called Earth-Prime. Eventually the Silver Age Justice League of America of the 1960s, consisting of the Flash, Green Lantern, the Atom, Batman, Superman, Wonder Woman, and other superheroes, met and had an adventure with Earth-2's Justice Society of America from the 1940s, whose membership contained the Flash, Green Lantern, the Atom, Batman, Superman, Wonder Woman, and others. So popular was this meeting of the two super-teams that it quickly became an annual tradition. But the Justice League and Justice Society could only visit each other's Earth so many times before the novelty wore off. Soon the Justice League branched out and visited other Earths, such as Earth-3, where the evil analog of the Justice League of America had formed the Crime Syndicate of America (presumably to distinguish themselves from their European criminal counterpart). Captain Marvel—that is, Billy Batson, who could become a superhero by shouting "Shazam!"—and the rest of his supporting cast inhabited Earth-S, and were in due time

---

* While Barry's explanation isn't actually all that exciting, it was the common practice in Silver Age comic books that every sentence that wasn't a question be punctuated with an exclamation mark!

paid a crossover visit by the Justice League of America.* Earth-X, Earth-4, and others soon followed, and eventually the phrase "multiverse" became appropriate to describe the seemingly endless number of alternate universes that abounded.

The issues of the *Justice League of America* that describe the meeting of the Silver and Golden Age heroes always carried titles such as "Crisis on Earth-2" or "Crisis on Earth-X." The story lines eventually became so convoluted, with so much alternate history to keep straight, that in 1985 DC Comics attempted to normalize the multiverse. The yearlong miniseries describing this simplification process was called "Crisis on Infinite Earths." With a vast housecleaning of continuity under way, the writers and editors at DC Comics used this opportunity to weed out poor sellers from many of the less-popular worlds and bring all the heroes from the better selling titles together on one Earth (coincidentally, the Earth-1 of the Silver Age heroes). Consequently the *Crisis on Infinite Earths* miniseries is noteworthy to comic-book fans for the deaths of the Barry Allen Flash and Supergirl (both of whom died heroic deaths struggling against the evil tyrant who threatened to destroy the Silver Age Earth-1) and the removal of Superboy from Superman's history. Unlike most comic-book fatalities, both Barry Allen and Supergirl have remained dead for the most part (with occasional resurrections as guest stars as a sales gimmick), while crimefighting adventures have begun to creep back into Clark Kent's teen years once again.

As ridiculous as all of the above may sound, the concept of an infinite number of parallel worlds may be one of the strangest examples of comic books getting their physics right![†] Just four years prior to the publication of *Showcase* # 4, the notion of an infinite number of parallel, divergent universes was seriously proposed as an interpretation of the equations of quantum mechanics. To reiterate: *Some*

---

* By this time DC had won its lengthy lawsuit against Fawcett, Captain Marvel's publisher, for copyright infringement on their Superman character. Contesting the lawsuit had contributed toward Fawcett's near bankruptcy, and DC was able to purchase the rights to the Captain Marvel character. Apparently, now that they owned the copyright to the character, DC was no longer concerned that publication of a Captain Marvel comic book would unfairly compete with and harm the economic viability of Superman comics.

† This time, the exclamation point is justified!

scientists believe that the concept of parallel universes is a serious, viable construct in theoretical physics. Current theories indicate that if such alternate Earths exist, they would be more like those described in the Marvel comic universe, where slight changes in a character's history (such as those presented by the Watcher in stories such as "What if Gwen Stacy Had Lived?") lead to diverging worlds that can never be visited by our reality, regardless of one's vibrational frequency.

GREAT MINDS REALLY DO THINK ALIKE

So far throughout this book we have discussed what physicists term "classical mechanics." To understand the "mechanics" of something means you can predict the statics (for example, the largest angle at which a ladder may be placed to remain balanced against a wall) and the motion of objects (such as the speed with which the ladder falls when it begins to slip), once the external forces acting on them are identified. The fundamental equation that governs how a macroscopic object will move as the result of an applied force is our old friend Newton's second law of motion, $\mathbf{F} = \mathbf{ma}$. For the motion of large objects such as cars, baseballs, and people, the dominant forces are gravity, friction, and electrostatics. Even when we considered electricity and magnetism, we still made use of $\mathbf{F} = \mathbf{ma}$, where the $\mathbf{F}$ in the left-hand side of this equation was either Coulomb's force of attraction or repulsion between electric charges or the force of a magnetic field on a moving electrical charge. The aspect of "quantum mechanics" that justifies its separation from "classical mechanics" as a distinct branch of physics is that when one considers electrons and atoms, $\mathbf{F} = \mathbf{ma}$ suddenly doesn't cut it anymore.

After a great deal of effort trying to "fix" classical physics for atoms (essentially tweaking Newton's laws without overturning them completely), physicists were reluctantly forced to conclude that a different type of "mechanics" applied inside of atoms. That is, a new equation was needed to describe how atoms would respond to external forces. After roughly twenty-five years of trying one form or another for this equation, nearly simultaneously Werner Heisenberg and Erwin Schrödinger obtained the correct form for an atom's equivalent of $\mathbf{F} = \mathbf{ma}$.

Have no fear: There is no chance that we will discuss either the Heisenberg or Schrödinger approach in any mathematical detail. In a few pages I'll write down the Schrödinger equation, but that will only be so that we can gawk at it as if it were some exotic zoo animal. Heisenberg's treatment of quantum physics involves linear algebra, while Schrödinger's makes use of a complex partial differential equation (for simplicity, we will focus on Schrödinger for the remainder of this chapter). To fully explain their theories would break the pact we have maintained up till now that nothing more elaborate than high school algebra would be employed in these pages (the algebra we have used in this book so far has the same relationship to linear algebra as a housefly does to a house).

Nevertheless, there are two points about mathematics worth making here. The first is that, unlike the case of Isaac Newton discussed in chapter 1, who had to *invent* calculus in order to apply his newly discovered laws of motion, both Heisenberg and Schrödinger were able to make use of mathematics that had already been developed at least a century earlier. The mathematical branches of linear algebra and partial differential equations that Heisenberg and Schrödinger employed to describe their physical ideas had been invented by mathematicians in the eighteenth and nineteenth centuries and were well established by the time they were needed in 1925.

Frequently, mathematicians will develop a new branch of mathematics or analysis simply for the pleasure of constructing a set of rules and discovering what conditions and principles then logically follow. Occasionally, physicists later discover that in order to describe the behavior of the natural world under investigation, the same tools that previously existed solely to satisfy mathematicians' intellectual curiosities turn out to be indispensable. For example, Einstein's task in developing a general theory of relativity in 1915 would have been much more difficult if he didn't have Bernhard Riemann's theory of curved geometry—developed in 1854—to work with. This scenario of physicists making tomorrow's advances by using yesterday's mathematical tools has been repeated so often that physicists tend to not think about it too much.

The second point about the theories of Heisenberg and Schrödinger is that, while they employ different branches of mathematics and look very different, upon close inspection (which Schrödinger performed in 1926) they can be shown to be mathematically equivalent. Because they describe the same physical phenomena (atoms, electrons, and light), and are motivated by the same experimental data, it is perhaps not very surprising that they turn out to be the same theory, even though the mathematical languages used to express them are vastly different.

Schrödinger and Heisenberg, completely independent of each other, developed different descriptions of the quantum world in the same year. The notion that some ideas become "ripe" for discovery at certain moments in history is found time and again and is not confined to theoretical physics. Of course, simple mimicry accounts for much of the similarity in television programs or Hollywood movies, just as the breakthrough of Superman in *Action Comics* led to a rapid proliferation of superhero comic books by many other publishers, including National Comics, hoping to bottle lightning again. There are, however, well-documented cases of movie studios or television networks simultaneously and independently deciding that it is time for the reintroduction of a particular genre, such as the pirate movie or the urban doctor drama. This synchronicity also occurs in comic books, as in the example of the X-Men and the Doom Patrol. In March 1964, DC Comics published *My Greatest Adventures* # 80, featuring the debut of a team of misfit superheroes (Robotman, Negative Man, and the obligatory female teammate, Elasti-Girl) whose freakish powers caused them to be shunned by normal society. They were led by a wheelchair-bound genius named the Chief, who convinced them to band together to help the very society that rejected them, frequently fighting their opposites in the Brotherhood of Evil. Three months later comic fans could buy *X-Men* # 1, published by Marvel Comics, where they would meet a team of mutants (Cyclops, Beast, Angel, Iceman, and the obligatory female teammate, Marvel Girl) whose freakish powers caused them to be shunned by normal society. These superpowered teens were led by the wheelchair-bound mutant telepath Professor X, who recruited and trained them to help the very society that rejected them, frequently fighting their opposites in the Brotherhood of Evil Mutants.

Despite some profound differences (Professor X is bald and clean-shaven, while the Chief has red hair and a beard), the striking similarities in concept have led many comic-book fans to wonder whether the X-Men were modeled after the Doom Patrol. However, interviews with the writers of both comics and the research of comic-book historians indicate that it is more likely that the near simultaneous appearance is a coincidence. The long lead time needed to conceive, write, draw, ink, and letter a comic book prior to its printing and newsstand distribution suggests that the X-Men were well into production by the time the Doom Patrol first appeared.

Another publishing synchronicity is the case of the moss-covered muck-monsters *Swamp Thing* at DC (written by Len Wein) and *Man-Thing* at Marvel (co-written by Gerry Conway), which appeared within one month of each other in 1971. Both Wein and Conway insist that their creations were not influenced by each other, and the fact that they were roommates at the time is purely coincidental.

* * *

If the behavior of objects on the atomic scale is governed by the matter-waves that accompany their motion, then what atomic physics needs is a matter-wave equation that describes how these waves evolve in space and time. Physicists in the early 1920s tried to construct such a matter-wave equation, until Erwin Schrödinger in 1925 (fig. 28) essentially guessed the right mathematical expression.

With the Schrödinger equation, scientists had a framework with which they could understand the interactions of atoms with light. This was Schrödinger's motivation for developing his matter-wave equation. A generation later, armed with the insights about the nature of matter made possible by the Schrödinger equation, a new class of scientists developed the transistor and, separately, the laser as well as nuclear fission (as in atomic bombs) and nuclear fusion (in hydrogen bombs). The path to both the transistor and the laser were difficult ones, and it was only with the guidance of quantum theory that they were successfully developed. A generation later still, the CD player, personal computer, cell phone, and DVD player—to name only a few—would be created. And because none of these are possible without the transistor or the laser, none of them are possible without the Schrödinger equation. It is small wonder that until fairly recently Schrödinger's portrait was on the 1,000 Schilling note in his native Austria, for he can truly be

$$-\hbar^2/2m \; \partial^2\Psi/\partial x^2 + V(x,\,t)\Psi = i\hbar\,\partial\Psi/\partial t$$

Fig. 28. *Erwin Schrödinger, theoretical physicist, Nobel Laureate, and ladies' man, thinking of the Schrödinger equation—the foundation of quantum mechanics and our modern technological lifestyle.*

considered one of the architects of the lifestyle that we in the twenty-first century take for granted.

A moment ago I said that Schrödinger "guessed" the form of the matter-wave equation. Perhaps "guessed" is too strong a word. Erwin Schrödinger used considerable physics intuition to develop a new equation to describe the behavior of atoms. Mere mortals may never know how exactly someone such as Newton or Schrödinger does what they do. The insight that leads to a new theory of nature is perhaps more powerful than that of artistic creation, for a new physical theory must not only be original but also mathematically coherent and agree with experimental observations. The most elegant theory in the world is useless if it is contradicted by experiments.

While we may not know how Schrödinger did what he did, we do know where and when he did it. Historians of science inform us that Schrödinger developed his famous expression in 1925 while staying at a Swiss Alpine chalet that he borrowed from a friend during a long Christmas holiday. Furthermore, while we know that his wife was not staying at this chalet, we also know that he was not alone. Unfortunately, we do not know which of Schrödinger's many girlfriends was with him at the time.

At this point the reader may wish to reexamine Schrödinger's portrait in figure 28. We may have a new answer to the question "Why is this man smiling?" Certainly Erwin does not strike one as a traditional ladies' man. If one ever wondered whether there might be some mathematical expression that would make one attractive to the opposite sex, the Schrödinger equation might be a good starting point. Even the brief overview of quantum physics presented in this chapter will no doubt, true believer, enhance your romantic desirability. This is in addition, of course, to the irresistible sex appeal that an encyclopedic knowledge of superhero comic books brings!

SCHRÖDINGER'S CAT OF TWO WORLDS

Schrödinger's equation is the $F = ma$ for electrons and atoms. Just as Newton's second law, once the external forces F are specified, describes the acceleration a, and from this the velocity and position of an object, Schrödinger's equation, given the potential energy of the electron through the term V, enables the calculation of the probability per volume ψ^2 of finding the electron at a certain point in space and time. Once I know the probability of where the electron will be, I can calculate the *average* location or momentum of the electron. Given that the *average* values are the only quantities that have any reliability, this is actually all that should be asked of a theory.

The emphasis on *average* quantities in quantum physics is different from our consideration of averages in the earlier discussion of thermodynamics (chapter 12). There we spoke of the average energy per atom in an object—characterized by its temperature—simply because it was convenient. In principle, if we had enough time and computer memory, or were super-fast like the Flash or Superman, we could keep track of the position and momentum of every air molecule in a room, for example. We could thereby calculate the instantaneous force on the walls per unit area, which would convey the same information as a determination of the average pressure. In quantum systems, on the other hand, the wavelike properties of matter set a limit on our ability to carry out measurements, and the average is as good as we'll ever get.

What is it about the wavelike nature of matter that makes it so difficult to accurately measure the precise location of an electron in an atom? Think of a clamped violin string with a fundamental vibration frequency and several higher harmonic tones. Assume that the string is vibrating at a given frequency, but one that we cannot hear. If the vibrations were so fast that we could not see the string vibrate back and forth, how would we verify that the string is indeed vibrating? One way would be to touch the string and feel the vibrations with our fingers. If our fingertips were sensitive enough (like Matt Murdock's, also known as Daredevil's), we could even determine the exact frequency that the string had been vibrating.

I say "had been vibrating," because once we have touched the string, it will no longer be oscillating at the same frequency as before. It will either have stopped shaking altogether or be vibrating at some different frequency. Perhaps we can determine the vibration frequency by bringing our fingers near to, but not in direct contact with, the string. In this way we can sense the vibrations in the air caused by the oscillating violin string. In order to improve the sensitivity of this measurement, we need to bring our fingertips very close to the string. But then the air vibrations will bounce from our fingers and ricochet to the string, providing a feedback that can alter its vibratory pattern. The farther away we hold our fingertips, the weaker the feedback, but then our determination of the vibratory frequency will be less accurate.

The matter-wave oscillations of an electron within an atom are just as sensitive to disturbances. Measurements of the location of an electron will perturb the matter-wave of the electron. Much has been written about the role of the "observer" in quantum physics, but it's no more profound than when you try to look at something smaller than the probe you are using to view it, you will disturb what you are trying to see.

Quantum theory can provide very precise determinations of the average time one must wait before half of a large quantity of nuclear isotopes has undergone radioactive decay (defined as the "half-life") but is not useful for predicting when a single atom will decay. The problem with single events is best illustrated by the following challenge: I take a quarter from my pocket and am allowed to flip it once and *only* once. What is the probability that it will come up heads? Most likely your gut instinct is to answer 50 percent, but you'd suspect a trick. And you'd be right—it *is* a trick question. To those who say the chance of getting heads is fifty-fifty, I say: prove it. And you can't, not based upon a single toss, as long as we live in world containing two-headed quarters. If you toss the coin a thousand times (or toss one thousand coins once) you would find that for a fair coin, it would land heads-side up very close to 50 percent of the time. But probability is a poor guide for single, isolated events. Yet probability is all that the Schrödinger equation offers. This did not sit well with many older physicists who were accostumed to the clockwork precision of Newtonian mechanics, and they proposed a conceptual experiment that would open a "Pandora's box," into which they placed a cat.

They posed the following situation: a box, inside of which is a cat and also a sealed bottle of poison and another, smaller box that contains a single radioactive isotope. The radioactive element has a half-life of one hour, which means, according to quantum mechanics, that after one hour, there is a fifty-fifty chance that it will have decayed. A by-product of this nuclear decay is the emission of an alpha particle (otherwise known as a helium nucleus), and the bottle of poison is arranged such that it will break open if struck by this particle. So, after one hour, there is a 50 percent chance that the cat is dead, having succumbed to the poison vapors released when the bottle was struck by the alpha particle, and a 50 percent chance that the bottle remained undisturbed, leaving the cat alive and well.

According to Schrödinger's equation, prior to the one-hour time limit, at which point one opens up the box and looks inside, the cat can only be meaningfully described as "the superposition (average) of a dead cat and a live cat." Once the lid is opened, the "average cat's wave function" collapses into one describing either a 100 percent live or 100 percent dead cat, but there is no way to know which will be observed before the lid is opened. If the walls of the box are transparent, you can never be sure that the light from the outside has not disturbed the decay process (recall that observing quantum systems can sometimes alter them). This interpretation has been found wanting by many physicists (despite the fact that recent experiments on entangled quantum states of light, as described in *JLA* # 19, the latest version of the Justice League of America, suggest that this is exactly what does occur), and a great deal of thought and argument has gone into attempting to resolve the intellectual unpleasantness associated with Schrödinger's cat. One provocative solution to this problem, described below, enables the Flash and Superman to travel to alternate Earths.

In 1957 Hugh Everett III argued that once the cat is sealed in the box, two nearly identical parallel universes are created and split off from each other: one in which at the end of the hour the cat is alive and another in which it is dead. What we do when we open up the box does not involve collapsing wave functions, nor is the cat 50 percent dead and 50 percent alive before we take a look. Rather, all we do at the end of the hour is determine which of the two universes we live in—one where the cat lives or one where the cat dies. In fact, for *every* quantum process for which there are at least two possible results, there are that many universes, corresponding to the different possible outcomes. Once the two Earths have been split off, due to the two possible outcomes of a particular quantum event, they each evolve in different ways, depending on the myriad further quantum events that occur following this initial branching point. If the bifurcation of the Earths occurred recently, then a particular Earth may be similar to our own world. If the separation occurred a long time ago, then during the intervening time there would be many opportunities for subsequent quantum events to have outcomes different from what was observed in our world. The history of this second Earth may be very much like ours then, but there is also the possibility of dramatic differences.*

Hence quantum theory provides a physical justification for both the "What If?" tales in the Marvel universe and the alternate Earths in DC comics. On one Earth Jay Garrick inhaled "hard water vapor" in a chemistry lab accident, gaining the gift of superspeed with which he fought for justice as the Flash with his teammates in the Justice Society of America. On another Earth police scientist Barry Allen was doused with an array of chemicals while

---

* An excellent recapitulation of the above discussion of the Schrödinger Cat thought-experiment can be found in *Animal Man* # 32. Though, for best effect, start at the beginning of this story arc with issue # 27.

simultaneously being struck by lightning, leaving him with the gift of super-speed with which he fought for justice as the Flash with his teammates in the Justice League of America. On another Earth a super-speedster committed crimes as the evil Johnny Quick with his teammates in the Crime Syndicate of America. There are in principle an infinite number of Earths, corresponding to all possible outcomes of all possible quantum effects, though a basic tenet of this theory is that ordinarily there can be no communication between these multiple alternate Earths. Ordinarily. Apparently for someone able to vibrate at super-speed like the Flash, travel between these many worlds could occur as often as readers kept buying such stories.

To physicists, Hugh Everett III's proposal led to a very different crisis of infinite Earths. The many-worlds solution to the Schrödinger's Cat problem represented to most physicists an example of the cure being worse than the disease. Nevertheless, there is nothing logically or physically inconsistent with this theory, and no one has been able to prove that it is incorrect.

Physicists who considered it intellectually unsatisfying to say that a complete theory of nature can only predict probabilities could not stomach the notion that the theory actually described the spontaneous and continuous creation of an infinite number of alternate universes. The "many-worlds" model has been considered the crazy aunt of quantum theory since its publication, and has been locked in the metaphoric attic until very recently. It was never taught to me, for example, when I studied quantum mechanics in college and again in more detail in graduate school. I discovered the "many-worlds" model completely by accident when, as a graduate student, I came across a copy of Bryce DeWitt and Neill Graham's 1973 book *The Many-Worlds Interpretation of Quantum Mechanics*, left abandoned in a graduate student office. In a successful attempt to procrastinate in doing my homework, I picked up this strange book, began reading it, and thereupon learned that somewhere there was another James Kakalios who was actually finishing his assignment on time (not that this knowledge did me any good).

Though few physicists give the "many-worlds" model the time of day, there is one class of theoretical physicists who have proven to be big supporters of the idea: string theorists.

WHY SUPERMAN CAN'T CHANGE HISTORY

In the years following the development of the Schrödinger equation, scientists have developed techniques to describe how the electron's matter-wave interacts with electric and magnetic fields (a process called Quantum Electro-Dynamics or QED) and how the matter-waves of quarks inside a nucleus behave (a process termed Quantum Chromo-Dynamics or QCD). A remaining goal of theoretical physics is to understand how matter-waves interact with gravitational fields. There is a perfectly good theory for gravity, namely Einstein's General Theory of Relativity. There is an excellent theory to describe the quantum nature of electrons (QED). Combining these theories into one coherent whole has proven beyond the abilities of any scientist living today. The closest that theorists have come to a quantum theory of gravity is something called "string theory."

A gross oversimplification of string theory is that it suggests that mass is itself a wave, or rather a vibration of an elemental string, and that these "strings" are the basic building blocks of everything in the universe. In its current state, many physicists are skeptical about string theory. Their first objection is that in order for the equations to balance, string theory only works in eleven dimensions (ten spatial and one time). This is somewhat awkward because, as near as we can tell, we only live in three spatial dimensions, and no one has ever encountered additional dimensions.* To address this discrepancy string theorists have suggested that there really are eleven dimensions, but seven of these spatial dimensions curl up into little balls, with a diameter less than a billionth trillionth trillionth of a centimeter, a length scale labeled the Planck length. Another drawback of string theory is tied to this extradimensional notion: Probing length scales so small requires correspondingly higher energies than current and next generation particle accelerators can acheive. Without the verification provided by experimentation, the only criterion to determine whether the equations are

---

* Except such imps as Mr. Mxyzptlk, Bat-Mite, Mopee, and possibly Dr. Strange when he was warned to "Beware Triboro! The Tyrant of the Sixth Dimension" in *Strange Tales* # 129.

on the right track is mathematical elegance. This may be dangerous; while it is true that the equations of classical mechanics, electricity and magnetism, and quantum mechanics do indeed have a certain mathematical beauty, there is no *a priori* reason to believe that nature really cares whether we find the equations pretty or not. Nevertheless, string theory is presently the only likely candidate for a quantum theory of gravity and only further study will determine its success.

Physicists developing quantum gravity have invoked the many-worlds interpretation in order to resolve logical inconsistencies in their calculations involving time travel. Recently some scientists have claimed that time travel is not physically impossible, though it is highly unlikely to ever be actually accomplished. The problem with time travel into the past is set forth in the famous "grandfather paradox." Essentially, if one could indeed travel back in time, it would be possible to murder your grandfather when he was a young man, before your own father was conceived. In this way you would prevent your own birth, but the only way you could have prevented it is if you had first been born. In order to find a way around this conundrum, modern theoretical physicists have dusted off Hugh Everett III's many-worlds interpretation. If there are indeed an infinite number of alternate parallel universes, then (the theorists argue) when you travel backward in time, the severe distortions in space-time necessary to make this journey would also simultaneously send you to a universe parallel to your own. You are therefore free to kill as many grandparents as you have bullets for, without fear of altering your own existence, because your own grandfather is safe in the past in your own universe, undisturbed by the havoc you are wreaking in alternate past worlds.

These modern theoretical notions were actually anticipated in the 1961 adventure "Superman's Greatest Feats" in *Superman* # 146. In this story Superman agrees to travel into the past as a favor for Lori Lemaris, a mermaid from the sunken city of Atlantis with whom he had a "special relationship" (while she was a girl and was his friend, Lori was not Superman's girlfriend). Lori beseeches Superman to prevent the sinking of Atlantis, which occurred millions of years ago. Superman argues that all of his previous attempts (presented in earlier issues of *Action Comics* and *Superman*) to change history have failed, but Lori's pleading (and what appear to be bedroom eyes) convinces Superman to give it a try. Given that it takes

Fig. 29. Superman travels through time and saves Abraham Lincoln from be-ing shot by John Wilkes Booth. Or does he?

great effort and a velocity larger than 1,100 feet per second to break the sound barrier (the effort, as discussed in chapter 5, is due in part to the work one must do to push the air out of his way), it was pro-posed in DC comics that with an even greater effort and a much faster velocity one could pass through the "time barrier." (Both the Flash and Superman, each capable of these necessary speeds, would travel back and forth through time as their story lines required.) Su-perman thus zooms to at least 8,000,000 B.C. and reaches nearly the exact moment when the advanced civilization of Atlantis, which resides on a small island off the shore of what appears to be a coastal resort, is about to succumb to "giant waves caused by a colossal un-dersea earthquake." Superman races to another island a safe dis-tance from the undersea quake, which is the home to *another* advanced civilization. Why we have never heard about this other ancient civilization is not addressed. Superman borrows some

Fig. 30. *Superman from the same story as fig. 29, now realizes that history has remained unchanged despite all of his time-traveling "Greatest Feats."*

"strange metal" from buildings about to be torn down on this other island and fashions an enormous crane by which he lifts the entire island of Atlantis, depositing it onto a secure deserted island, where it is spared by the earthquake. (Let's not even get into what this "strange metal" could possibly be composed of, that it would have a tensile strength sufficient to lift an island.)

This time, unlike all previous attempts, Superman *was* able to successfully change the course of history, and he decides to make various pit stops on his return trip to his own time period, using this opportunity to "fix" various historical events. He saves the Christians from being eaten by lions in the Roman Coliseum,*

---

* The story uses the variant spelling "colosseum." Actually the correct spelling for the location where Christians were devoured in ancient Rome is: Circus Maximus.

Fig. 31. *Supeman in 1961 discovers what quantum theorists have only recently hypothesized—that travel through time must of necessity also involve transport to alternate, parallel universes.*

takes Nathan Hale's place as he is about to be executed by the British, prevents Custer's massacre at Little Big Horn, and drops by Ford's Theater on April 14, 1865. As shown in fig. 29, as John Wilkes Booth is about to assassinate President Lincoln, he has time to utter "*Sic Semper*—Ulp!" as hands that can crush diamonds close upon his pistol. Superman is now like a kid in a historical candy store, and decides to try to save the population of his home planet Krypton. Because he loses his superpowers under the red light of Krypton's sun Rao (by this time the explanation for his amazing abilities had been attributed to Earth's yellow sun) Superman decides to build a fleet of spaceships from sunken Earth naval vessels and send them to Krypton in order to enable everyone to escape to another world. Using his telescopic vision, he watches his parents disembark on a new planet with an infant Kal-El. At this point Superman realizes that he has stumbled upon a paradox, for if his parents never sent him to Earth as a baby, how is he able to save them now?

Returning to his present in 1961, the Man of Tomorrow discovers that all of the history books are unchanged, shown in fig. 30. Lincoln was indeed shot at Ford's Theater and Nathan Hale and General Custer are similarly described suffering their Superman-less fates. Superman can't understand how this can be, since "surely, the [history] books are truthful." Ahem. Retracing his path through the time stream, Superman comes across an alternate Earth (fig. 31), in which the history books give proper credit to Superman's role in correcting the past's "mistakes."

Alas, Superman discovered in 1961 what theoretical physicists have rediscovered in 2001—that time travel is only possible via the many-worlds interpretation of quantum mechanics. Superman did indeed accomplish these amazing feats, altering the course of history—but in an alternate universe, not in his own (see fig. 31). A similar phenomenon occurs in the Marvel comics *Avengers* # 267 where the evil time lord Kang the Conqueror is revealed to have single-handedly created a vast number of alternate Earths as a by-product of his frequent time traveling in order to defeat his superhero foes. Still another example of comic books being ahead of the physics curve.

THROUGH A WALL LIGHTLY—

TUNNELING PHENOMENA

ONE ASPECT OF QUANTUM MECHANICS that is difficult for budding young scientists to accept is that the equation proposed by Schrödinger predicts that under certain conditions matter can pass through what should be an impenetrable barrier. In this way quantum mechanics informs us that electrons are a lot like Kitty Pryde of the X-Men, who possesses the mutant ability to walk through solid walls (as shown in fig. 32), or the Flash, who is able to "vibrate" through barriers. (illustrated in fig. 33). This very strange prediction is no less weird for being true.

Schrödinger's equation enables one to calculate the probability of the electron moving from one region of space to another even if common sense tells you that the electron should never be able to make this transition. Imagine that you are on an open-air handball court with a chain-link fence on three sides of the court and a concrete wall along the fourth side. On the other side of the concrete wall is another identical open-air court, also surrounded by a fence on three sides and sharing the concrete wall with the first court. You are free to wander anywhere you'd like within the first court, but lacking superpowers you cannot leap over the concrete wall to go to the second court. If one solves the Schrödinger equation for this situation, one finds something rather surprising: The calculation finds that you have a very high probability of being in the first open-air court (no surprise there) and a small but nonzero probability of winding up on the other side of the wall in the second open-air court (Huh?). Ordinarily the probability of passing through a barrier is very small, but only situations for which the probability

Fig. 32. A scene from X-Men # 130, *showing Kitty Pryde (not yet a member of the X-Men) as she employs her mutant ability to walk through walls to sneak up on the White Queen of the Hellfire Club.*

is exactly zero can be called impossible. Everything else is just unlikely.

This is an intrinsically quantum mechanical phenomenon, in that classically there is no possible way to ever find yourself in the second court. This quantum process is called "tunneling," which is a misnomer, as you do not create a tunnel as you go through the wall. There is no hole left behind, nor have you gone under the wall or over it. If you were to now run at the wall in the other direction it would be as formidable a barrier as when you were in the first open-air court, and you would now have the same very small

Fig. 33. Scene from Flash # 123, *where Jay Garrick, the Golden Age Flash, demonstrates the quantum mechanical process known as "tunneling." The matter-wave of an object has a small but nonzero chance of passing through a solid barrier. The faster the object approaches the barrier, the greater the transmission probability. As Jay correctly notes, the barrier is unaffected by the tunneling process.*

probability of returning to the first court. But "tunneling" is the term that physicists use to describe this phenomenon. The faster you run at the wall, the larger the probability you will wind up on the other side, though you are not moving so quickly that you leap over the wall. This is no doubt how the Flash, both the Golden and Silver Age versions, is able to use his great speed to pass through solid objects, as shown in fig. 33. He is able to increase his kinetic energy to the point where the probability, from the Schrödinger equation, of passing through the wall becomes nearly certain.

Consider two metals separated by a vacuum. An electron in the metal on the left is like a person in the first open-air handball court. Instead of a concrete wall, a thin vacuum separates this electron from the second metal that can be considered another open-air court. An electron in one metal has a small but nonzero probability of finding itself in the second metal. The electron does not arc across the vacuum gap and does not have enough kinetic energy to escape from the metal on its own. (This is a good thing. Otherwise all objects would be continually leaking electrons all over the place, and static cling would be one the most gripping problems of the day.) Rather, the electron's matter-wave extends into the gap, decreasing in magnitude. A similar phenomenon occurs with light waves moving from a denser to a less dense medium. Under conditions for which the light wave should be totally reflected at the interface, there is still a small diffraction of light into the less dense medium. The diffracted wave's magnitude decreases the further into the less dense medium it progresses. Since the square of the electron's wavefunction represents the probability of finding the particle at a point in space and time, a finite magnitude for the "matter-wave" indicates that there is a probability the electron is in the second metal. If the gap is not too large (compared to the electron's matter wavelength, which in practice means roughly less than one nanometer), then the matter-wave will still have an appreciable magnitude in the second metal. Let us be clear, the electron on one side of the barrier moves toward the obstruction, and most times simply reflects off the wall. If a million electrons strike the barrier then, depending on its height and width, 990,000 electrons might be reflected and 10,000 would wind up on the other side.

If the separation between the two metals is too large, then even for the most energetic electrons the chance of tunneling becomes very, very small. A person's momentum is large, so our matter-wavelengths are very small—much less than a trillionth trillionth of the width of an atom and much smaller than the width of the concrete wall separating us from the second open-air handball court. Nevertheless, if you were to run toward the concrete wall, there is a very, very small probability that your matter-wave will arrive on the other side of the wall. The greater your kinetic energy, the larger your chance of tunneling. Those who doubt that this is indeed possible are invited to begin throwing themselves at concrete walls right now, and to persevere in their attempts no matter how discouraging the initial results.

Electrons in a solid rattle around at a rate of more than a thousand trillion times per second. Consequently, in one second they have a thousand trillion opportunities to tunnel through a barrier. Send enough electrons against a barrier, and if the height of the wall is not too high or the thickness of the separation too large, an appreciable fraction will indeed tunnel to the other side. Not only has the phenomenon of quantum-mechanical tunneling been verified for electrons, but it is the central principle behind a unique type of microscope called a Scanning Tunneling Microscope that enables one to directly image atoms. As shown in fig. 34, when a

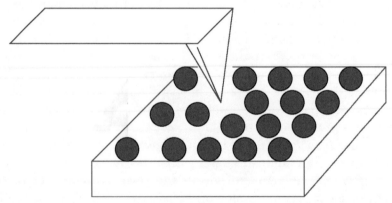

Fig. 34. *Cartoon showing the basic mechanism of a scanning tunneling microscope. A fine metal tip is brought very close to a conducting surface. By very close I mean within a few atoms' widths. When the tip passes over an atom on the surface, the electron probability clouds of the atom may lead to quantum-mechanical tunneling into the tip. When the tip is right above an atom, the tunneling probability is high, and the current in the tip is large. In this way the atoms of the surface can be scanned and imaged.*

metal tip is brought very close to, but not touching, a metal surface, it can intercept the electron clouds surrounding each atom on the surface. When the electrons tunnel from an atom to the metal tip, an electrical current is recorded in a meter connected to the tip. Whether or not tunneling occurs is very sensitive to the separation between the atoms on the surface and the scanning metal tip. A change in distance of just the width of an atom can change the probability of tunneling by a factor of more than a thousand. By moving the tip slowly over the surface and carefully measuring the current at each location, the position of each atom on the surface can be mapped out.

Just such an image is shown in fig. 35, which shows the location of carbon atoms on the surface of a crystal of graphite (more commonly known as "pencil lead"). The gray scale is not real (carbon atoms aren't really black or white, or any color for that matter) but is used to represent the magnitude of the current recorded in the tip at any position, which in turn reflects the electron density at each spot. Fig. 35 shows us that the carbon atoms in graphite form hexagons much like the six-sided plates that make up a snowflake. The fact that the carbon atoms form a hexagonal lattice implies that a crystal of graphite consists of sheets of carbon atoms as in fig. 35 lying atop one another. If you pressed snowflakes together, the hexagonal plates would lock in place much

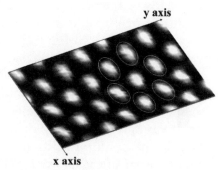

Fig. 35. *Scanning Tunneling Microscope image of the surface atoms of graphite, the form of carbon used in pencil lead. Each white spot indicates a region of space where the tunneling current was high for that tip location (see fig. 34). The hexagonal lattice of carbon atoms is readily apparent. The grayscale is used to indicate the intensity of the tunneling current. The y-axis extends for 1 nm while the x-axis is 0.5 nm long. Courtesy of Dr. Laura Adams and Prof. Allen Goldman at the University of Minnesota.*

like the carbon atoms in graphite. Building a three-dimensional crystal out of such two-dimensional sheets, the solid essentially stacks each sheet atop the other like the thin layers in a puff pastry. The planes in solid graphite are so loosely held together that you can easily peel them apart just with your hand, simply by scraping a pencil point along a sheet of paper. The fact that this form of solid carbon makes a better writing implement than if all carbon atoms had four equally strong bonds (otherwise known as "diamond") can be inferred directly from this atomic image.

In the next chapter we will discuss the physics of transistors and diodes, and I'll give away the punchline now and tell you that these semiconductor devices are essentially valves that regulate and amplify the flow of current. One way this current can be controlled is through the tunneling process. When two conductors are set close to each other, separated by a thin insulating barrier, normally no current can flow from one conductor to the other. By applying a voltage across this sandwich structure, the effective height of the wall separating the electrons of one region from the other can be varied. As noted, the tunneling probability is a very sensitive function of this barrier height. In this way the tunneling effect is used to modulate the flow of electrons across the device. These "tunneling diodes" are integral components of cell phones, as well as many other solid-state devices. Quantum-mechanical tunneling is therefore not an esoteric theoretical novelty or useful solely in atomic microscopes. Many of the products we associate with our current lifestyle would not be possible without tunneling being a reliable phenomenon.

When we apply the laws of quantum physics to large objects like Kitty Pryde of the X-Men (fig. 32), we find that tunneling is still possible, but very unlikely. How unlikely? Assuming Kitty's mass is 50 kilograms (such that her weight is 110 pounds— one of her code names was Sprite, after all), then even if she could run at a wall as fast as she could, a million times a second, it would take longer than the age of the universe before she could expect to quantum-mechanically tunnel through to the other side. Clearly the one-time miracle exception comes into play in a big way here. With our improved understanding of physics, we can now more accurately describe Kitty Pryde's mutant power as being able to alter her macroscopic quantum wave function, increasing her tunneling probability to near 100 percent at will. Quite useful when one has locked the keys inside the car.

A long-standing puzzle in comic books is, if Kitty Pryde can walk through walls, why doesn't she fall through the floor at the same time? How, when she is "phasing" and immaterial, can she walk? In *X-Men* # 141, it was argued that while phasing, Kitty actually walks on a surface of air, and is not in actual contact with the floor. While immaterial in her phasing mode she is therefore unaffected by any trapdoors opening beneath her. Assuming for the moment that she can indeed walk on air—that is, that somehow the air provides enough resistance to the backward force of her feet that it supplies a forward thrust, propelling her forward— the question still arises as to how her partially material foot can follow her body through a wall.

However, if the mechanism by which she is able to pass through solid barriers is indeed quantum-mechanical tunneling, then it is perfectly reasonable that she would not slide through the floor. When an electron tunnels from one side of a barrier to the other, it conserves energy in the process. If it has a certain energy on one side of the barrier, it has the same total energy after the tunneling process is complete. In fact, tunneling can only occur when the energy of the object is exactly the same on both sides of the barrier. In which case, if she were to speed up by falling through a trapdoor while tunneling, where would she gain this extra kinetic energy from? Similarly, she could never slow down, as she would have to make contact with her surroundings in order to transfer some of her kinetic energy.

Technically, she cannot walk while tunneling, as she may not increase her energy by pushing against any object, whether the object is the solid floor or a cushion of air. But at the same time, she cannot lose any energy either. All she needs to do is walk normally as she approaches a wall, turn on her mutant power of maximizing her tunneling probability, and she will glide through the partition with the same speed as she had when she neared it. For those times when she desires to phase through a floor, such as in *Astonishing X-Men* # 4, where she actually phases through nearly a hundred feet of solid metal to reach an underground laboratory, she would jump up slightly while in her corporeal mode, and then right before her feet touch the ground, activate her mutant tunneling ability. She would continue her motion with the last kinetic energy she had while solid, and descend with a steady velocity. It's probably safer for her if she keeps her mutant tunneling power activated until she is near the floor in the lower room, and avoids becoming material near the ceiling, where she would have to deal with her now large gravitational potential energy.

23

SOCK IT TO SHELLHEAD–

SOLID-STATE PHYSICS

IF ANY SUPERHERO demonstrates the value of a physics education, and in particular the study of semiconductors that is part of solid-state physics, it is Iron Man, who wore a high-tech suit of armor to fight for justice. The heart of Iron Man's amazing offensive and defensive capabilities was a modern (in 1963) technological marvel: the transistor.

The transistor truly was a revolutionary device, as its ability to amplify and modulate voltages had a profound impact on all of our lives. Initially the transistor was used solely to duplicate the functions of vacuum tubes, so that radios and television sets could become lighter and more efficient. As scientists and engineers developed techniques for making transistors smaller and smaller, their use for mathematical calculations led to the development of electronic computers. The solid-state transistor is the fountainhead of nearly every electronic device in use today. At this point we have enough physics under our belt to understand how this remarkable device works. Before delving into semiconductor physics, let's review the events that led to Shellhead's* comic-book debut.

The Cold War loomed large in the monthly adventures of Silver Age comic books of the late 1950s and early 1960s. In DC comics, fighter pilots with the "right stuff" figured prominently in several offerings. When an alien Green Lantern crash-landed on Earth and

---

* The cylindrical helmet of his original suit gave Iron Man the nickname "Shellhead," which has clung to him for the past forty years, regardless of how streamlined and stylish later versions of his helmet became.

was at death's door in *Showcase* # 22, he instructed his power ring
to seek out someone brave, honest, and fearless to whom he could
bequeath the ring and power lantern. The ring selected an American test pilot, Hal Jordan. (In DC's *Showcase* # 6 another fighter
pilot, Ace Morgan, led the Challengers of the Unknown.) The
Marvel Age of comics began in 1961 (the same year that both Russian and American astronauts first traveled to space), when four
adventurers—a scientist, his girlfriend, her teenage brother, and a
former fighter pilot—took an unauthorized rocket flight through a
cosmic radiation belt in order to beat the Communists "to the
stars." The cosmic rays they absorbed would turn this quartet into
the Fantastic Four.

The "red menace" of Communism continued to raise its head
in Marvel superhero comic books fairly often. The Hulk, who appeared a year after the debut of the Fantastic Four, owed his existence to Commie spies, when physicist Dr. Robert Bruce Banner
was exposed to an overdose of gamma radiation from a detonat-
ing Gamma Bomb. His assistant (who was in reality a Communist spy—you would think that a research assistant named Igor
would have been a tip-off when he received his security clear-
ance, but never mind), deliberately did not stop the countdown,
hoping to eliminate America's leading bomb expert, and creating
the Hulk instead.

The Soviet presence in early Marvel comics was faced by the
Human Torch, Ant-Man, Spider-Man, the Mighty Thor, and the
Avengers. But none of the mighty Marvel superheroes of the early
1960s were so closely allied with the Cold War as the invincible
Iron Man. *Tales of Suspense* # 39 introduced the brilliant inventor
and industrialist Tony Stark to the Marvel universe. Using his
mastery of transistorized technology, Stark developed many new
weapons for the military as part of his effort to help the United
States win the fight over Communism in Indochina. Not content
to merely test his new weapons in the lab, Stark accompanies an
inspection team into the jungles of Vietnam, so that he could more
accurately assess the effectiveness of his inventions. Alas, we soon
see why more CEOs do not take such a hands-on approach to quality control. A booby trap kills the military advisors traveling with
Stark, and leaves Tony himself with a piece of metal shrapnel
lodged in his chest, dangerously close to his heart. To make matters
worse, he is captured and brought to the hidden camp of Wong-Chu,

"the red guerrilla tyrant." A doctor at Wong-Chu's camp determines that the shrapnel is actually migrating and that in a few days it will reach Stark's heart and kill him.

Wong-Chu offers Stark a deal: Work in my weapons development lab (this is apparently a pretty well-equipped guerrilla camp) in return for surgery that will save your life. Stark agrees, intending to use his brief remaining days to create some sort of weapon to both save his life and combat his captor. Along with the brilliant physicist Professor Yinsen, also a captive of Wong-Chu, Stark constructs a metal chestplate that, once fully electrically charged, will prevent the shrapnel from reaching his heart. Realizing that he will need offensive and defensive weapons if he and Yinsen are to escape the guerrilla camp, the chestplate becomes part of an iron suit, which contains an array of transistorized weaponry. The suit is completed moments before the shrapnel can claim Stark's life, though Prof. Yinsen is killed as the chestplate finishes charging up. Yinsen's death is avenged, and the other prisoners in the camp are freed as the Iron Man defeats the Communist warlord. Making his way back to the States through the jungles of North Vietnam (a story not fully told until years later in *Iron Man* # 144), Tony Stark would continue to use his technological suit of armor to defend America against Communist aggression.

And boy did Shellhead attract the Communist foes! At times he seemed to have a "red magnetism" field built into his suit of armor. Borne out of the Vietnamese conflict, Iron Man would battle more Communist villains in his first four years than nearly all other Marvel superheroes combined. Iron Man would face the Red Barbarian (*Tales of Suspense* # 42); the Crimson Dynamo (*Tales of Suspense* # 46 and 52), a Russian power station built in the form of a suit of armor designed to defeat Iron Man; the Mandarin (*Tales of Suspense* # 50, 54, 55, 61, 62), a Chinese warlord modeled after the Sax Rohmer pulp-fiction villain Fu Manchu, who possessed ten deadly rings of power; and the Titanium Man (*Tales of Suspense* # 69–71), a Soviet-built stronger version of Iron Man created to defeat our hero in televised conflict, thereby proving to the world the superiority of communism over capitalism. Throughout all this, Tony Stark maintained the fiction that Iron Man was a separate person, hired by Stark to serve as his bodyguard. Given the number of times Communist agents tried to kidnap Stark or steal his research plans, this was not such a far-fetched cover story.

As if the constant battles with costumed villains were not distracting enough, Stark was continually called to testify before Senate committees, who insisted that it was his patriotic duty to turn the Iron Man technology over to the military. Little did Senator Byrd (*not* the long-serving senator from West Virginia), who led the investigations into the connections between Iron Man and Stark Industries, realize, the secret of Iron Man's success—the transistor—was public knowledge. The transistor was developed by three physicists in 1947 at Bell Laboratories in Murray Hill, New Jersey—the research lab of utility Bell Telephone. In order to facilitate the adoption of this new technology, Bell Labs ran seminars for other firms interested in using transistors, instructing them in the details of the new field of solid-state physics. It's not enough to build a better mousetrap; you must also make sure that the mice know about it!

CLOTHES MAKE THE MAN

Following his first appearance, Iron Man's suit would undergo nearly constant modifications, both cosmetic and significant. The suit was gray in *Tales of Suspense* # 39, but by the very next issue Stark decided to change the color to gold, so as to make more of an impression on women. You would think that a multimillionaire industrialist who looked like Errol Flynn (the model upon which artists based their drawing of the mustachioed Tony Stark) would not have to worry about whether his secret identity as Iron Man was attractive to the ladies, but it is presumably just such attention to detail that brought Tony Stark his success. Within a year the suit would be redesigned again, now as a more form-fitting yellow-and-red ensemble shown in fig. 36 that, with minor variations, would persist to recent times.

The weapons that were distributed throughout the suit would also undergo near-constant upgrades. Initially Stark had "reverse magnetism" projectors in the palms of his gloves, but these were soon modified to "repulsor rays" that were essentially "force beams." A large recessed disc on his chest housed a "variable power spot-light" that evolved into a "uni-beam" (I'm not really sure what this did). He originally had a radio antenna extending from his left shoulder, but improvements in wireless transmission and reception technology enabled this function to be incorporated within the body of his iron suit.

The armor itself is still quite heavy, even in its more flexible, streamlined form. The only way that Stark is able to walk in this iron suit, and to lift objects weighing up to several tons, is through the application of "tiny transistors within his armor that increase his power tremendously." And he's going to need this extra power. In order to estimate how much the armor weighs, assume that the Iron Man suit is one eighth of an inch thick and has the average density of iron, which is roughly 8 grams per cubic centimeter.

© 1963 Marvel Comics

Fig. 36. *Panels from a bonus section "All about Iron Man" in* Tales of Suspense *# 55, providing a schematic of the 1960s Golden Avenger, and tutorial stating that the charge to power supplies feeding his transistors is depleted the more he uses them.*

The surface area needed to make the suit can be approximated by taking Tony's trunk as a cylinder, his head as another, smaller cylinder, and his arms and legs as smaller, longer cylinders (remember the spherical chicken story in chapter 10? This time we'll assume a cylindrical Iron Man). If Tony is about six feet tall and his suit jacket is a 50 Regular, then his total surface area is roughly 26,200 square centimeters. The volume of iron within his suit is found by multiplying the surface area by the armor's thickness of 1/8 inch, or 0.32 centimeters, yielding a total volume of metal of roughly 8,400 cubic centimeters. To determine how much it weighs, we multiply this volume by the density (8 grams/cm^3) and obtain 68,000 grams or 148 pounds, excluding the weight of all of the transistorized circuitry. Tony Stark would frequently carry all of this in his briefcase (with one dress shirt covering the armor as camouflage—see *Tales of Suspense* # 55), except for the chestplate, of course, which he wore constantly in order to keep the shrapnel from worming its way to his heart. Consequently, simply by lugging nearly 90 pounds of armor around in this briefcase, Tony would have developed considerable upper-body and bicep strength as a side benefit of being Iron Man.

The suit's weight leads to the question of how his jet boots enable Iron Man to fly. If the suit weighs nearly 150 pounds, and Stark himself tips the scales at 180 pounds, then his boot thrusters have to supply a downward force of 330 pounds (equal to a mass of 150 kg), just for Iron Man to hover in the air. Presumably these jets use a chemical reaction to violently expel the reactants from the soles of his boots. As every action is balanced by an equal and opposite reaction, this downward force leads to an upward push on Iron Man, keeping him aloft. If he wants to accelerate, then his boots have to provide even more force, given that only the force in excess of his weight will provide an acceleration ($\mathbf{F = ma}$).

Tony would frequently need to travel from his Stark Industries plant on Long Island to Avengers Mansion in the heart of Manhattan, a distance of approximately fifty miles (as the Golden Avenger flies) in a time of ten minutes. This corresponds to an average speed of 300 mph, which is nearly half the speed of sound! Ignoring the energy that Shellhead must expend to push the air out of the way as he travels at this speed, this implies that for a kinetic energy $(1/2)\mathbf{m}\mathbf{v}^2$ at this speed, his armor requires at least 1.37 million kg-meters^2/sec^2 of energy. In comparison, the average person expends over eight million kg-meter^2/sec^2 of energy in a full day. When Iron Man needed to travel great distances he would eschew the boot jets and make use of motorized roller skates that were built into his boots. Not only would it be more fuel efficient, since energy does not have to be expended counteracting gravity in keeping Shellhead in the air, but whenever he is decelerating he can use the rotational energy and an alternator to recharge his internal batteries, as in an automobile. In this way Tony Stark anticipated recent hybrid-engine automotive technology.

In the seventies Iron Man had gone green, and his armor was now coated with a thin layer of solar cells, enabling him to recharge whenever he was in direct light. The energy from the sun on an average day in the United States is roughly 200 kg-meter^2/sec^2 per second over an area of 1 meter^2. We have just calculated the surface area of the Iron Man suit to be 26,200 centimeters^2, which means that the amount of energy striking Iron Man per second is 262 kg-meter^2/sec^2 (at any given moment only half of his available surface area can be facing the sun) while his boots require an expenditure of power of more than a million kg-meter^2/sec^2. If the solar cells are 50 percent efficient in converting the energy of the sun into stored energy in Tony's battery packs (and most commercially available solar cells have a conversion efficiency of only about 10 percent), Iron Man would need nearly three hours to soak up enough sunlight for this one trip. We have considered neither the energy needed to run the suit's internal air-conditioning unit (pushing air out of the way at 300 mph will make any person inside a metal suit a tad sweaty), nor if he has to fire his repulsor rays during the flight. In the normal course of a typical day in the life of Iron Man, he will expend energy much faster than he can recharge his storage batteries using solar cells.

To give the writers of the Iron Man comics credit, their concern with mechanisms for Tony to recharge his armor's storage batteries implies a recognition of the principle of conservation of energy. From his very first appearance in *Tales of Suspense* # 39, it was always acknowledged that running the mechanized suit Stark developed required large amounts of energy and that the greater his expenditure of power, the faster the drain on the energy reserves he might carry on his person. Not only was a ready supply of electrical energy necessary to animate his jet boots and activate the servo-motors that enabled him to move in the suit and increased his strength, but his chestplate needed electrical energy as well in order to protect his heart from the shrapnel he carried with him ever since that fateful day in Vietnam. The 1960s Iron Man would occasionally have to drag himself dramatically along the ground after a particularly energy-exhausting battle, searching for an electrical outlet in order to recharge his battery reserves.

Even after he made the transition to solar-powered battle armor, Stark's suit could run dry in an emergency. In *Iron Man* # 132, Tony drained every last erg of energy (one erg is one ten millionth of a kilogram-meters^2/sec^2) from his suit in an exciting, no-holds-barred battle with the Incredible Hulk. Tony focused all of his suit's stored energy into one final punch, and accomplished what had previously been impossible: Iron Man knocked the Hulk unconscious. But the cost to Tony Stark was high. With absolutely no power to move his suit, Stark was trapped, unable to move within his now rigid shell of armor. To make matters worse, the protective covering over his eye and mouth slits had been engaged, to shield Stark from the exploding jet. Tony was therefore facing suffocation once the air contained in the suit was used up. It would take all of the following issue for Ant-Man, forcing his way in through the exhaust port in Iron Man's boot jet, to travel the length of the armor, avoiding the suit's internal protective mechanisms, and disengage the faceplate's protective covering.

HE FIGHTS AND FIGHTS WITH REPULSOR RAYS

Of all of Iron Man's weapons, his most effective are his "repulsor rays," which are emitted from discs on the palms of his armored gloves. Back in his first appearance in *Tales of Suspense* # 39, the first version of this glove-based repulsion weapon was a "reverse magnetism" ray used to fight his way out of Wong Chu's prison camp. Wong Chu's guards, finding that small-arms fire bounced harmlessly off the iron suit that the intruder wore, responded by preparing to shoot bazookas and throw grenades at the Yankee invader. Fig. 37 shows that as they fetch the heavy weaponry, Tony takes the time to "reverse the charge on this magnetic turbo-insulator and use a top-hat transistor to increase its repelling power a thousandfold!" As the rays are emitted from his hand, deflecting the weapons, he exclaims, "There! Reverse magnetism—it

Fig. 37. *Iron Man, in his first appearance in* Tales of Suspense *# 39, fights his way out of a Vietnamese prison camp using a top-hat transistor and a "magnetic turbo-insulator."*

works like a charm!" In fact, it would have to work like a charm, because there's no way it could work using solid-state physics.

There's only one aspect of the scene summarized above that is physically correct, and that involves the "top-hat transistor." There is no such thing as a "magnetic turbo-insulator"; this is just techno-babble. The "turbo" modifier is just to make these insulators sound cool. There are magnets that are nonmetallic—that is, they are electrical insulators yet still generate a large magnetic field, and devices called "top-hat transistors" do indeed exist. They are so named because they look like small cylinders, about the size of pencil erasers (this was back in the early 1960s, long before the microminiaturization of transistors enabled millions of such devices to be fabricated on a chip measuring only a few millimeters on each side), with a small disc at their base at which the electrodes extended, looking a little like the top-hat playing piece from a Monopoly game set. The panel showing Tony Stark employing such a device to amplify the current to his "magnetic turbo-insulator" is physically plausible. But the second-to-last panel, in which he then employs said device to deflect the grenades and bazooka shots using "reverse magnetism" is not.

While every electron, proton, and neutron inside every atom has an intrinsic magnetic field, the natural tendency of magnets to line up, north pole to south pole, has the effect of canceling out the magnetism of most atoms. Any magnetic field Iron Man would create using a powerful electromagnet in the palm of his glove would only be effective if: (1) the grenades being tossed at him were for some reason already magnetized and (2) they were all perfectly thrown so that their north poles were all pointing in the same direction and (3) the magnetic field created by Iron Man's hand was also oriented so that the north pole was directed toward the incoming grenades and not the south pole, which would have the effect of accelerating the weapons toward him. It is unlikely that Tony Stark could always count on his opponents to cooperate with suitably oriented magnetic weapons.

Ironically, Iron Man's reverse-magnetism ray has a better chance of working on nonmagnetic objects! Recall our discussion in chapter 18 concerning Magneto and the phenomenon of diamagnetic levitation. Unlike metals such as iron or cobalt, for which the internal atomic magnetic fields align in the same direction, many materials, including water, are diamagnetic. In this case when they are in an external magnetic field, the atomic magnets orient themselves to oppose the applied field. In this case all of the south poles of the atomic magnets line up to point in the direction of the external magnet's south pole—that is, opposite to the way an iron magnet would behave. Thus the very process of trying to magnetize the object leads to a repulsive force. In chapter 18 we saw that if the magnetic field created by Magneto is more than 200,000 times larger than the Earth's magnetic field, then this repulsive force can overcome the downward weight of the object, lifting it up off the ground. Similarly, Iron Man's reverse magnetism could repel objects, but only if they were diamagnetic, and it would not work on many metallic objects that are either ferromagnetic or paramagnetic (which align with an applied field). Magneto creates these large magnetic fields through his mutant power, but Iron Man must do it the old-fashioned way, using electromagnets (similar to the one constructed by Superboy in chapter 18). As Iron Man does not carry an electrical dynamo around with him, a few shots of this reverse-magnetism ray would drain his batteries faster than a fight with the Hulk. Furthermore, the recoil of such weapons is considerable. When supplying a large force striking their target, they will induce an equal and opposite force to the gun and the shooter holding it. Tony Stark was clever to build his repulsor rays into his gloves. By locking the servo-motors that enable his armored arms to move, his iron suit provides a large and rigid inertial mass to take up the recoil whenever he fires this glove-based weapon.

While "reverse magnetism" may not be physically practical, hand-held pulsed-energy weapons have begun to make the transition from comic-book fantasies to military research facilities. Certainly these weapons cannot be the same "magnetic repulsors" as Iron Man uses, for the reasons argued above. The energy needed to generate a magnetic field large enough to deflect an object using only diamagnetic repulsion is so large that it would be more effective to employ conventional weaponry. Nevertheless, "pulsed" energy systems are under active development by the military. By generating a large voltage inside the weapon that can be rapidly discharged in a thousandth of a second, the power (energy divided by time) could be quite high. This electromagnetic pulse, if directed at a target, would deposit this energy in a localized region faster than the heat could be safely dissipated away. High-intensity laser beams delivered in extremely brief pulses are used in physics laboratories to nearly instantaneously melt a small region of a crystal's surface, and in principle the same process could be employed in an offensive capability. The big drawback is the energy requirements of such a weapon. If one must carry a miniature power plant around in order to fire such a pulsed energy weapon, the element of surprise in any combat situation would be lost.

SOLID-STATE PHYSICS MADE EASY

What is a transistor, this piece of electronics that, according to Stan Lee at least, is endowed with miraculous abilities that enable Iron Man to successfully fend off the Mandarin, the Crimson Dynamo, and Titanium Man? A short answer is that transistors are valves that regulate the flow of electrical current through a circuit. Such answers are easy to remember, but they tell us nothing about how transistors actually function. The first question we should address is: What exactly is a semiconductor, that is neither a metal nor an insulator? We hear a great deal about how we are living in the "Silicon Age," but what is so special about silicon? In the next few pages I will try to condense more than fifty years of solid-state physics as I answer these questions.

Silicon is an atom, a basic element of nature, just like carbon, oxygen, or gold. A silicon atom's nucleus has 14 positively charged protons and (usually) 14 electrically neutral neutrons, and to maintain charge neutrality there are 14 negatively charged electrons surrounding the nucleus. These electrons reside in the "quantum-mechanical orbits" that, as discussed in chapters 20 and 21, arise from the wavelike nature of all matter. The possible "electron orbits" are specific for each element, and determine the allowed electron energies.

Quantum mechanics enables us to calculate, via the Schrödinger equation, the allowed "orbits" of the electrons in an atom, and knowing how many different possible orbits an electron can have in an atom is like knowing the number and arrangement of chairs in a classroom (stay with me here; this classroom metaphor is going to be useful in explaining metals, insulators, and semiconductors). The chairs only represent possible or virtual classes; it is not until the students enter and take their seats that the class is real. If only one student comes in and takes a seat, this is like having only one electron in a possible quantum-mechanical orbit. We would call this class Hydrogen, in analogy with the atom that has only one electron in its neutral, stable form. If there were two students sitting in the class, we would have Helium, fourteen students would make up Silicon, and so on. The first students to enter the class take the seats at the front of the room, close to the blackboard in our hypothetical example. The last students to enter take seats near the back of the auditorium, far from the blackboard (where the positively charged nucleus will be). This arrangement, with every seat filled by a student, describes the lowest-energy configuration. For a carbon atom with six electrons, the closest orbits are occupied. If the carbon atom gains some energy, say, from absorbing light, some of its electrons will then occupy higher-energy orbits.

Whether a material is a metal, a semiconductor, or an insulator depends on the energy separation between the highest level filled with an electron and the nearest available unoccupied level. In the classroom analogy the solid can be thought of as a very large auditorium with many rows of seats, provided by the constituent atoms that make up the material. There will be an empty balcony that contains an equal number of seats. If the electrons sitting in the lower-energy orchestra seats* are to conduct electricity when a voltage is applied across the solid, then they gain extra energy. They can only absorb this energy if there is an empty state at higher energy for the electron to move into (recall the discussion of quantized energy levels in chapter 20). The electrical properties of any solid are determined by the number of electrons residing in the lower orchestra seats and the energy separation between the lower occupied seats and the next empty ones in the balcony.

The difference between insulators and metals is clear in this analogy. An insulator is a solid where every single seat in the orchestra is filled, while a metal is a material for which only half of the seats in the lower level are occupied. In a metal there are a large number of empty seats in the orchestra available to an electron, and the application of a voltage, whether big or small, can accelerate the electrons to higher energy states (which correspond to carrying an electrical current). Metals are good electrical conductors because their lowest occupied auditorium seats for electrons are only half-filled. For the insulator, every seat is occupied, and absent promotion to the balcony, no current will result when a

---

* The lowest-energy levels are the first to be filled with electrons. Strictly speaking, every pair of electrons in an atom gets its own auditorium (they pair up due to their intrinsic magnetic fields, north pole to south pole). It is the last electrons to get placed in available levels that determine the chemical reactivity and electronic properties of the corresponding solid, and it is the auditorium containing these electrons that we consider here.

voltage is set up across the material. If I raise the temperature of an insulator, providing external excess energy in the form of heat, some of the electrons can rise into the previously empty balcony. In the balcony there are many empty seats for the electron to carry a current, but this will last only as long as the temperature is elevated. If the temperature is lowered, the electrons in the balcony will descend and return to their low-energy seats in the orchestra.

If the insulator absorbs energy in the form of light, it can immediately promote an electron to the balcony. When the electron returns to its seat in the orchestra, it has to conserve energy and thus gives off the same amount of energy that it previously absorbed. It will either do this by giving off light of the same energy as initially absorbed, or the electron can induce atomic vibrations (heat). This is why shining light on an object warms it up—the electrons absorb the energy of the light, but then can return the absorbed energy in the form of heat. If the energy of the light is insufficient to promote an electron from the highest filled orchestra seat to the lowest empty balcony seat, the light is not absorbed. In this case the lower-energy light is ignored by the electrons in the solid, and passes right through it. Insulators such as window glass are transparent because the separation between the filled orchestra and empty balcony for this material is in the ultraviolet portion of the spectrum, so visible light with a lower energy passes right through. On the other hand, metals always have available empty seats to absorb light even in the half-filled orchestra. No matter how small the light's energy, an electron in a metal can absorb this energy and then return it upon going back to its lower-energy seat. This is why metals are shiny. They always give off light energy equal to what is absorbed, and there is no lower limit to the energy of light they can take in.

A semiconductor is just an insulator with a relatively small energy gap (compared to the energy of visible light) separating the filled lower band from the next empty band. For such an energy separation, a certain fraction of electrons will have enough thermal energy at room temperature to be promoted to the balcony. When electrons are excited to the upper deck, the material now has two ways to conduct electricity. For every electron promoted to the higher energy band that is able to conduct electricity, an empty state is left behind. The empty chairs in the previously filled orchestra can be considered as "positive electrons" or "holes," and can also carry electrical current. If an electron adjacent to an empty seat slips into this chair, then the empty spot has migrated one position over. In this way we can consider the hole to move in response to an external voltage, and also carry current. Of course, the original electrons will eventually fall back down into the orchestra, filling the empty seats they left behind (though not necessarily their original seats). When certain semiconductors absorb light, there are enough excited electrons in the upper band and holes in the lower band to convert the material from an insulator to a good electrical conductor. As soon as the light is turned off, the electrons and holes recombine, and the material becomes an insulator again. These semiconductors are called "photoconductors" and are used as light sensors, as their ability to carry an electrical current changes dramatically when exposed to light. Certain smoke detectors, television remote controls, and automatic door openers in supermarkets make use of photoconductors for their operation.

Semiconductor devices are typically constructed out of silicon because it has an energy gap conveniently just below the range of visible light. Furthermore, it is a plentiful element (most sand is composed of silicon dioxide) that is relatively easy to purify and manipulate. There are times when the physical constraints of the size of the energy gap in silicon limits a device's performance, and in this case other semiconducting materials can be used, such as Germanium or gallium arsenide. Iron Man's, and the military's, night-vision capabilities make use of a semiconductor's photoconducting properties and a small energy gap that is in the infrared portion of the electromagnetic spectrum.

All objects give off electromagnetic radiation due to the fact that they are at a certain temperature, so their atoms oscillate at a particular frequency that reflects their average kinetic energy. On a dark, moonless night, the temperature of most nonliving objects decreases (as they are not absorbing sunlight), so they emit less radiation and at lower frequencies. Humans, on the other hand, have metabolic processes that maintain a uniform temperature of 98.6 degrees Fahrenheit. Consequently, we emit a fair amount of light (as much energy as a 100 watt lightbulb) in the infrared portion of the spectrum. Our eyes are not sensitive to this part of the spectrum, but semiconductors can be chosen that have a large photoconductivity when exposed to infrared light. At night the infrared light given off by a warm-blooded person is much greater than his or her colder surroundings.

Certain night-vision goggles using "thermal imaging" detect this light by using semiconductors, which absorb the infrared radiation given off by an object at a temperature of roughly 100 degrees. The photocurrent in the semiconductor detector is then transported to an adjacent material, which is chemically constructed to give off a flash of light when the photoexcited electrons and holes recombine. In this way the infrared light that our eyes cannot usually detect is shifted to the visible portion of the electromagnetic spectrum, thereby enabling us to see in the dark. These goggles also detect visible light, as well as infrared light during the daytime. All objects give off roughly the same intensity of light if they are at the same temperature (recall our discussion of light-curves from chapter 20). When the objects around a person are warmer (due to absorbed sunlight), the contrast between the infrared light from a person and his or her inanimate surroundings is diminished, as is the utility of the goggles.

WHAT COLOR ARE THE INVISIBLE WOMAN'S EYES?

An understanding of semiconductor photoconductivity also helps to resolve a question that has long perplexed comic-book fans: Why isn't the Invisible Woman blind? When the Fantastic Four took their ill-fated rocket trip, Sue Storm (now Susan Richards) gained the ability to become completely transparent at will. How can she do this, and how can she see if visible light passes right through her? The more basic question is: How do we see anything at all?

The molecules that make up the cells in our bodies absorb light in the visible portion of the electromagnetic spectrum. The addition of certain molecules, such as melanin, can increase this absorption, darkening the skin. As a result of her exposure to cosmic rays, the Invisible Woman gained the ability to increase the "energy gap" of all of the molecules in her body (this is presumably the nature of her "miracle exception"). If the separation between the filled lower orchestra and the empty upper balcony is increased such that it extends into the ultraviolet portion of the spectrum, then visible light will be ignored by the molecules in her body and pass right through her. This is not so far-fetched; after all, we all possess invisible cells that are transparent to visible light. In fact, you're using them right now, reading this text through the transparent lens of your eyes.

Sunlight contains a great deal of ultraviolet light, which has more energy than visible light. We typically don't think about the ultraviolet portion of the solar spectrum until we get a sunburn on a bright summer day. When Sue becomes invisible, she still absorbs and reflects light in the ultraviolet region of the spectrum. We can't see her because the rods and cones in our eyes do not resonantly absorb ultraviolet light. Special UV glasses (like the ones Doctor Doom installed in his armored mask) could shift the ultraviolet light reflected from Sue down into the visible portion of the spectrum, using a similar mechanism to the one used by "night-vision" goggles in shifting low-energy infrared light up into the visible portion of the spectrum.

This also explains how the Invisible Woman is able to see. The rods and cones in her eyes, when she is transparent, become sensitive to the scattered ultraviolet light that bounces off us, and is ignored by our eyes. The world Sue sees while invisible will not have the normal coloring we experience, for the shift in wavelengths of the light she detects is not associated with the colors of the rainbow. Windows appear transparent to us because they transmit visible light and absorb ultraviolet light. We can't see ultraviolet light, so we don't notice its absorption. However, when Sue is invisible a window will appear as a large dark space while other objects will appear transparent to her. With a little practice she would be able to maneuver just fine.

This mechanism to account for Sue's ability to see while invisible was suggested in *Fantastic Four* # 62, vol. 3 (Dec. 2002) which corresponds to the 491st issue in the numbering scheme that began in 1961. In this issue, written by Mark Waid and drawn by Mike Wieringo, we are told that while invisible, Sue sees by detecting the scattered cosmic rays that are all around us but cannot be detected by normal vision. Right idea—wrong illumination source. Cosmic rays from outer space are not light photons but are mostly high-velocity protons that, upon striking atoms in the atmosphere, generate a shower of electrons, gamma ray photons, muons (elementary particles related to electrons), and other elementary particles. We usually don't have to worry about radiation damage or gaining superpowers via cosmic-ray-induced mutation, at least at sea level, as the flux of high energy particles is a million trillion times less than that of sunlight. If Sue depended on cosmic rays to see at street level she would be constantly bumping into objects and people. It is more likely that her vision makes use of the same mechanism by which she becomes transparent—namely, a shift in her molecular bonding into the ultraviolet portion of the spectrum.

WHAT IS A TRANSISTOR, AND WHY SHOULD WE CARE?

Back to Tony Stark and his transistorized suit of armor. When Tony needed to increase the repelling power of his magnetic turbo-insulator, he used a top-hat transistor. How are transistors able to amplify weak signals, making radios portable and repulsor rays powerful?

While semiconductors are useful as photoconducting devices, if this were their only application no one would think to call this era the Silicon Age. The thing about semiconductors that makes them very handy to have around the house is that you can change their ability to conduct electricity by a factor of more than a million just by intentionally adding a very small amount of chemical impurities. Not only that, but depending on the particular impurity, you can either add excess electrons to the semiconductor or remove electrons from the filled auditorium, thereby creating additional holes that can also conduct electricity. When a material with excess electrons is placed next to a semiconductor with additional holes, you have a solar cell, and if you then add a third layer with excess electrons on top of that, you've made a transistor.

It's been known for a long time that the addition of certain chemicals can change the optical and electronic properties of insulators. After all, that's how stained glass is made. Ordinary window glass has an energy gap that is larger than the energy of visible light, which is why it is transparent. But add a small amount of manganese to the glass when it is molten—and, after cooling, the glass appears violet when light passes through it. Manganese has a resonant absorption right in the middle of the glass's energy gap, as if we had parked some extra chairs on the stairways that connect the filled orchestra and the empty balcony. Particular wavelengths of visible light that would ordinarily pass through the material unmolested will now induce a transition in the magnesium atoms added to the glass. In this way certain wavelengths are removed from the white light transmitted through the glass, giving the window material a color or "stain." Different chemical impurities, such as cobalt or selenium, will add different colorations (blue and red, respectively) to the normally transparent insulator.

The same principle works for semiconductors, only the chemical impurities that we choose to add can either make it very easy to promote electrons to the balcony or to take electrons out of the filled auditorium, leaving holes in their place. A semiconductor for which the chemical impurities donate electrons is termed "n-type," since the electrons are negatively charged, while those for which the impurities accept electrons from the filled lower states are called "p-type," referring to the positively charged holes created. What's special about such semiconductors with added impurities is not that their conductivity can be changed dramatically (if we wanted a more conductive material, we would just use a metal) but rather what happens when we put an n-type semiconductor next to a p-type semiconductor. The extra electrons and holes near the interface between these two different materials quickly recombine, but the chemical impurities, which also have an electrical charge, remain behind. The positively charged impurities in the n-type region and the negatively charged impurities in the p-type region create an electric field, just as exists between positive and negative charges in space. This electric field points in one direction. If I try to pass a current through the interface between the n-type and p-type semiconductors, it will move very easily in the direction of the field, and it will be very tough going opposing the field. Such a simple device is called a "diode" in the dark, and a "solar cell" when you shine light on it. When the p–n junction absorbs light, the light-induced electrons and holes create a current, even without being connected to a battery. The charges are pushed by the internal electric field just as surely as if the device were connected to an external voltage source. A solar cell therefore can generate an electrical current through the combination of the newly light-induced extra electrons and holes with the internal electric field left behind by the charged impurities. This is one of the very few ways to generate electricity that does not involve moving a wire through a magnetic field, and thus no fossil fuels need be consumed for this device to work.

A transistor takes the directionality of the electrical current of a diode and makes the internal electric field changeable. By doing this, the transistor can be viewed as a special type of valve, where an input signal determines how far the valve is opened, which in turn leads to either a large or small current flowing through the device. Returning to the water flow analogy for electrical current from chapter 16, a fire hose is attached to the city water supply and, as the valve connecting the hose to the faucet is opened, water flows through the hose. If the valve is barely cracked open, the flow will be very weak, and as the valve is opened wider and wider, the quantity of water exiting the hose increases. Usually I have to manually turn the handle of the valve to effect a change. Now imagine a valve that is connected to a second, smaller hose that brings in a small stream of water. How much or how little the valve is opened will depend on how much water the second hose brings to the valve. If I considered the water flow in this second hose as my "signal," then the resulting water flow out of the main fire hose would be an amplified version of this signal.

In this way a small voltage can be magnified without changing any of the time-dependent information encoded within it. When Iron Man needs to increase the current to his magnetic turbo-insulator a thousandfold, or amplify the current going to the servo-motors that drive the punching force of his suit, he uses transistors to take small input currents and increase their amplitude. Despite what Tony Stark would tell you, transistors don't actually provide power, but they do enable the amplification of a small signal, increasing it many times. To do so they need a large reservoir of electrical charge, such as an external battery, just as in the water analogy the "transistor valve" would not amplify the weak input unless the fire hose was connected to the city water supply. Consequently, rather than providing power, transistors actually use power, but the rate at which they use power in order to amplify a weak signal is much less than the old amplification technique (vacuum tubes) they replaced. This is why Iron Man would be in desperate need of a recharge after a taxing battle. Tony would frequently gasp that his transistors needed to recharge, but I'm sure that he actually meant to refer to the battery supply to his transistors. Such a slip of the tongue is forgivable—I'm sure I'd misspeak after going several rounds with the Titanium Man.

Before transistors, the amplification of a weak input current was performed by heated wires and grids that guided the motion of electrons across space. A current was run through a filament wire until it glowed white-hot, and electrons were ejected from the metal and accelerated by a positive voltage applied to a plate some distance away, pulling these free electrons toward it. Between the filament and the collector plate is a grid (that is, a screen) which can act like a valve. If the input signal was applied to this grid, it would modulate the collected current, opening and closing the valve as in the water analogy. In order to avoid collisions with air molecules that would scatter the electron beam away from the collector electrode, these wires and grids were enclosed in a glass cylinder from which nearly all the air had been removed. These so-called vacuum tubes were large, used a great deal of power to heat the wires and run the collector plate, took a while to warm up when initially started, and were very fragile. Semiconductor-based transistors are small, low-power devices that are instantly available to amplify current and are compact and rugged. Even so, it took years before the transistor, invented in 1947, replaced the vacuum tube in most electronic devices.

One does not accidentally discover the transistor device, but must carefully construct a semiconductor structure with high purity and low defect density so that the amplification process can be observed. The hard work and innovative experimental techniques that enabled John Bardeen, Walter Brittain, and William Shlockley at Bell Laboratories in Murray Hill, New Jersey, to construct the world's first transistor, was recognized by the their Nobel Prize in Physics, awarded in 1956. On the day Bardeen learned that he had been awarded his second Nobel Prize (in 1972, for his development of a theory for superconductivity), his transistorized garage-door opener malfunctioned, underscoring the need for continuing research in solid-state physics.

As manufacturing- and quality-control techniques improved, and newer and smaller designs for transistors became available, another important application for this special electron valve was realized. With a small input current applied to the transistor, a small output current results. A relatively modest increase in the input current creates in turn an amplified, larger current. The output of the transistor can be either a "low current" or a "high current," and a low current is termed a "zero," whereas if there is a high current, then this state is labeled a "one." Minor adjustments to the inputs to a transistor can create either a one or zero for the output current. By combining literally millions of transistors in clever configurations, and making use of a branch of mathematics called Boolean logic (developed by a mathematician named George Boole more than 90 years before the transistor was invented and 70 years before Schrödinger's equation was developed), one has the basic building block of a microcomputer.

A full discussion of how computers manipulate "ones" and "zeroes" to represent larger numbers and carry out mathematical operations through binary code would require another, separate book. The point I want to make here is that at the heart of all microcomputers and integrated circuits is the transistor. The "chips" that underlie the commercial and recreational electronics that play a larger and larger role in society, from cell phones to laptop computers to DVD players, are all simply platforms for the clever arrangement of, and connections between, a large number of transistors. The computerized and wireless technology that surrounds us in the twenty-first century would not be possible without the transistor, which in turn could not have been invented without the insights previously gained by the pioneers of quantum physics and electromagnetism.

Schrödinger was not trying to develop a CD player, or even replace the vacuum tube, when he developed his famous equation, but without his and others' investigations into the properties of matter, the modern lifestyle we enjoy today would not be possible. All of our lives would be very different if not for the efforts of a relatively small handful of physicists, studying the behavior of the natural world. With few exceptions, these scientists were driven not by a desire to create commercial devices and practical applications, but rather by their curiosity, leading them, as Dr. Henry Pym put it in *Tales to Astonish* # 27, to "work only on things that appeal to [their] imagination."

SECTION 4

WHAT HAVE WE LEARNED?

ME AM BIZARRO!–

SUPERHERO BLOOPERS

WE STARTED THIS BOOK with a discussion of how Superman, applying Newton's laws of motion, can leap a tall building in a single bound, and ended with Kitty Pryde quantum-mechanically tunneling through solid walls and Iron Man's transistorized armor. Along the way we have addressed many of the major subjects that would be covered in a basic undergraduate physics curriculum, from the first topics addressed in introductory physics (such as Newton's laws of motion and the principle of conservation of energy) to upper-level material (quantum mechanics and solid-state physics). However, I would be remiss if I left you, true believer, with the impression that absolutely *everything* in superhero comic books is fully consistent with the laws of physics. I therefore would like to conclude by discussing some of those very few, rare examples where comic books actually get their physics *wrong*, no matter how many miracle exceptions one is willing to grant.

CYCLOPS OF THE X-MEN'S SECOND MUTANT POWER

The first young mutant that Prof. Charles Xavier recruited to join his nascent super-team the X-Men was Scott Summers, code named Cyclops. Scott's mutant gift, and also his curse, was that beams of "pure force" were emitted from his eyes. These force beams could punch a hole through a concrete wall and deflect a falling two-ton boulder. Only two materials were immune to Scott's optic beams:

his own skin (which was a good thing, or else his eye blasts would blow his eyelids off his face) and "ruby quartz." Scott is forced to either constantly wear sunglasses made of this exotic material—or a wraparound visor when he was on superhero duty. He could raise this ruby quartz shield using buttons either on the side of the visor or in the palms of his gloves. When he wore the visor, his eye blasts were projected as a single broad red beam of force, from which his nom de superhero was derived. When the ruby quartz shield was lowered, Scott could see the (red-tinged) world clearly, with the visor or sunglasses safely absorbing the destructive brunt of his optic force beams.

Quartz is the name that geologists have ascribed to the crystalline form of silicon dioxide. If the silicon dioxide molecules are arranged in a disordered fashion, as in a large number of marbles randomly poured into a container, then the resulting material is called "glass," but if the molecular units are carefully stacked in an ordered array, the mineral is called "quartz." Just as there are different ways that a collection of marbles can be stacked in a regular pattern, there are different crystalline configurations of quartz. If the mineral contains a very small amount of iron and titanium, the resulting crystal will have a slight pinkish hue (as in our discussion of stained glass in chapter 23), in which case it is called "rose quartz." A suspension of ruby dots in the quartz will result in cloudy brown and beige veins, and this dark, smoky, difficult to see through mineral is termed "ruby quartz."

As strange and inconvenient as it would be to have force beams projecting from your eyes, such that you would always have to look at the world through ruby quartz glasses, we cannot really protest, for no matter how physically implausible this may be, it is covered by our "one-time miracle exemption" policy. However, regardless of the mechanism by which Cyclops's optic blasts work, there is a key scene that is always missing in the X-Men comics and motion pictures whenever Scott lets loose with his mutant power. What we never see, yet know must occur by the laws of physics described earlier, is Scott's head snapping backward due to the recoil of his force beams.

Newton's third law informs us that forces always come in pairs—that is, every action is accompanied by an equal and opposite reaction. You cannot push against something if there is nothing to push against. Rockets depend on this principle when they expel hot gases at high velocities, so through Newton's third law the recoil propels the ship in the opposite direction of the exhaust. Similarly, a large beam of force, sufficient to keep a two-ton boulder suspended in midair (from which we conclude that the force of the optic blasts must be at least 4,000 pounds), should push back on Cyclops's head with an equivalent recoil force of 4,000 pounds. From Newton's second law of motion—that is, **Force equals mass times acceleration**—his body (assuming a mass of 80 kilograms) would rapidly acquire an acceleration of more than 20 times that of gravity. From an initial stationary position, his head should be moving backward at several hundred mph whenever he makes use of his special gift. Therefore we must conclude that in addition to powerful optic blasts Cyclops possesses a second, hidden mutant talent—namely, he is also endowed with exceptionally strong neck muscles.

PUT THAT BUILDING DOWN!

As mentioned in the beginning of this book, in the early days of the Golden Age of comics, Superman's powers were attributed to the fact that his home planet Krypton had a much stronger gravitational pull than Earth's. By using the benchmark that he is able to leap a tall building in a single bound on Earth, we calculated in chapter 1 that the acceleration due to gravity on Krypton had to have been at least fifteen times greater than our own. The Man of Steel was therefore not really made of metal, but had muscles and a skeletal structure adapted to a much larger gravity. Imagine lifting a full gallon container of milk, which weighs nearly nine pounds. If you want to experience what life would be like on a planet with a gravity fifteen times weaker than Earth's, you should empty the gallon container and refill it with just a little more than a half pint of milk. Compared to hefting the full gallon, you would find the same container with only eight ounces of milk to be much easier to pick up. Similarly in *Action Comics* # 1 Superman is able to raise an automobile weighing roughly 3,000 pounds over his head. A 3,000-pound weight to Superman (adapted to Krypton's heavier gravity) is similar to us lifting a 200-pound weight overhead.

We noted earlier, that with his rising popularity, Superman

morphed from being the champion of the little guy to the star of a multimillion-dollar marketing empire. The threats that Superman faced became more extreme and his foes became more super-powered. Big Blue's strength level correspondingly increased to fantastic levels. Before long he was lifting tanks, trucks, locomotives, ocean liners, jumbo jets, and high-rise office buildings. Similarly, Marvel Comics's hero the Incredible Hulk possessed a strength that also strained credulity. The Hulk's strength is tied to his bloodstream's adrenaline levels, which is why emotional stress triggers his transformation from puny Bruce Banner into eight feet of thickly muscled green rage. The correlation with adrenaline also accounts for the fact that the madder the Hulk gets, the stronger he becomes. When suitably aggravated, he has been known to pick up and throw a castle; the side of a cliff, and even hold up a mountain threatening to crush him and a collection of other Marvel super-heroes in the *Secret Wars* miniseries. While Reed Richards races to modify Iron Man's armor to channel both the Human Torch's nova-flame and Captain Marvel's electromagnetic energy in order to blast an escape tunnel out of the mountain, Reed deliberately insults and taunts the Hulk, knowing that their survival depends on a suitably steamed Jade Giant.

Eventually Superman would become so strong that, as shown in fig. 38 from *World's Finest* # 86, he could carry *two* high-rise office buildings—one in each hand—as if he were carrying two pizza pies, while flying at the same time! An examination of this figure reveals one reason he is able to cart off these buildings from Gotham City to an open-air exhibit in Metropolis: They were not connected to any city water or electrical utilities. As astounding as this display of strength is Superman's comment: "I got permission to borrow the two Gotham City buildings you asked for." As to exactly whom you would ask for verbal "permission" to carry off two high-rise buildings, I could not even begin to speculate. I doubt that either of those buildings' superintendents has the authority to allow Superman to borrow their building. But it's hard to say no when Superman asks if he could pick up your office tower and fly it to another city for a charity event. Better to just evacuate the building of all its workers and staff and say, "OK, Superman!"

Even if you accept that any person, whether a strange visitor from another planet or a nuclear scientist accidentally bombarded

I GOT PERMISSION TO BORROW THE TWO GOTHAM CITY BUILDINGS YOU ASKED FOR!

SPLENDID-- JUST SET THEM DOWN HERE AND WE'LL HAVE ONE OF OUR ASSISTANTS SHOW HOW THE *BURGLAR-MACHINE* WORKED!

AN A
SUR

Fig. 38. Panel from World's Finest # 86, *where the Man of Steel demonstrates a much-improved strength level compared to his first appearance in* Action # 1, *where he lifted an auto-mobile over his head, provoking the startled onlookers to flee in panic. In contrast, no one in the amphitheater is overly perturbed that a flying man is carrying two high-rise office build-ings over their heads.*

with gamma radiation, could indeed be strong enough to pick up a building, there is a separate violation of physics principles associated with these scenes: Simply put, buildings, ocean liners, and jumbo jets are not designed to be picked up. They are either intended to remain stationary, such as an office building, or to be supported at several points, such as the three wheels under an airplane on the runway, or, in the case of a battleship, uniformly buoyed up by the water they displace. The problem with lifting an office tower, for example, is that any slight deviation from the vertical will result in gravity creating an unbalanced torque, trying to twist the building even further toward the horizontal.

Buildings such as high-rise towers or castles are large, so the distance from the edge of the building to its center of mass is very long (termed the "moment arm" in chapter 8). These structures are quite heavy, so there is significant weight trying to rotate the building. The bigger the object, the larger the distance from its edge to the point where Superman or the Hulk is holding it, and the larger the moment arm of the torque trying to twist it. The torque for the buildings carried by Superman in fig. 38 is many times greater than reinforced concrete (concrete with steel rods threaded through it to increase its rigidity) can withstand before fracturing. Realistically, if you picked up a building and flew it somewhere, there would be a continuous stream of construction debris left behind you. Superman should arrive at the Gotham City charity event holding a few cinder blocks in each hand, and not two office towers with their structural integrity intact. Rather than asking for permission to borrow the buildings, Superman should practice asking for forgiveness for destroying these towers by picking them up in the first place.

Some later comic-book writers have realized that it's impossible, regardless of your level of super-strength, to pick up a building and not have it crumble in your hands. In Marvel's *Fantastic Four* # 249 a Superman stand-in, code named Gladiator, picks up the edge of the Baxter Building (the FF's headquarters) at its base and rocks it back and forth, but without physically harming the tower. Reed Richards, the smartest man in the Marvel universe, instantly recognizes that what Gladiator is doing is impossible. He theorizes that Gladiator must actually possess a previously unnamed superpower of tactile kinesis, defined in comic books as the ability to levitate an object that one is in physical contact with. Of course, there is no such thing as tactile kinesis, but it does reduce down to a manageable number the quantity of miracle exceptions that are needed for the story to progress.

If we compare Giant-Man to a redwood tree (and not just because his personality was sometimes a little stiff), we note that the taller the tree, the wider the trunk. In order to provide support for the large mass above it, a tree needs a very broad base. Around the time of the signing of the American Declaration of Independence, two mathematicians, Euler and LaGrange, proved that a column shorter than a certain height is stable, and will be compressed by the weight of material pressing down on its base, but above a certain height (whose value depends on the strength of the material comprising the column), the tower becomes unstable against bending. The slightest perturbation away from an exactly vertical orientation leads to a large twisting force, that is, a "torque" as in the case of the seesaw in chapter 8, that will cause the column to bend under its own weight. Giant-Man could, in principle, grow as tall as a redwood tree, but he would have to be just as mobile (assuming he stayed below the height limit set by the cube-square law—see chapter 10). Any attempt to run after or fight a supervillain would inevitably lead to the upper portion of his body leaning forward over his legs. The weight of his upper trunk would then cause his body to rotate, and before you could say "Stan Lee," old Highpockets would be flat on the ground.

Just such a fate inevitably befell Stilt-Man, an early foe of Daredevil's. Stilt-Man possessed a mechanized suit that contained two hydraulic legs that, when fully extended, enabled him to stand several stories tall. As sure as summer follows spring, Daredevil would use the cable in his billy club to tangle up Stilt-Man's legs, and the resulting loss of stability would bring the issue's adventure to a rapid close.

Another mystery related to the center of mass is how Spider-Man's foe, Doctor Octopus, is able to walk. Research scientist Otto Octavius employed four robotic arms that were attached to a harness around his waist, with which he manipulated radioactive isotopes. The inevitable explosive radioactive accident caused this harness and the arms to be fused to Octavius, and Doctor Octopus was born. But these arms are very heavy, and we frequently see him standing on his two legs while all four arms move behind him! They should therefore create a large torque that would put Doc Ock flat on his back, or on his face if they're in front of him. Spidey should be able to disable (if not disarm) Doctor Octopus by simply tossing an apple at him whenever he spots the arms not anchoring him to the ground.

The quick and unsatisfying resolution of these stories when the physics is taken too seriously should make abundantly clear why there is not a great demand for physics professors to write superhero comic books.

THE JUSTICE LEAGUE HAS THE MOON ON A STRING

Another unrealistic feat of strength occurs in the conclusion of a 2001 adventure of the Justice League (by this time they had dropped the "of America" part of their team name, though the comic featuring their adventures still went by the acronym *JLA*). In *JLA* # 58, Superman, Wonder Woman, and Green Lantern are shown pulling the moon into Earth's atmosphere in order to defeat a group of renegade Martians. Perhaps I had better back up and explain why they considered this a good idea.

Martians were introduced into the DC universe in 1955's *Detective Comics* # 225, when a physics professor trying to develop an interstellar communication device accidentally created a transporter beam instead. He thereby forcefully brought J'onn J'onzz (the Martian Manhunter) to Earth. J'onn eventually adopted a costumed identity as a superhero crime-fighter, and was a founding member of the Justice League of America back in 1960. J'onn J'onzz possessed a dazzling array of superpowers that matched Superman's—including flight, super-strength, invulnerability, Martian-breath (equivalent to Superman's super-breath), super-hearing, Martian vision, and several that Superman could only dream of, such as mental telepathy, invisibility, and shape-shifting. Just as Superman needed Kryptonite to keep him from always solving every problem in a nanosecond, the Martian Manhunter, being more powerful, required an even more common Achilles' heel in order to justify why he would ever bother teaming up and forming a league with other superheroes. It was therefore revealed that J'onn suffered, as did all Martians, from a vulnerability to fire. Consequently, rather than having to search for an exotic meteorite from the doomed planet Krypton, all you would need is a penny book of matches to incapacitate the Martian Manhunter.

It is revealed in the pages of *JLA* that J'onn J'onzz is mistaken when he considers himself the last survivor of the Martian race, when the Earth is attacked by a small army of evil Martians, each one possessing J'onn's superpowers. The Justice League lures the evil Martians to the moon, where the Martians have no fear of a fire weakening them. However, while J'onn J'onzz uses his mental telepathy to distract these villains, Superman, Wonder Woman, and Green Lantern employ an enormous cable to drag the moon into the Earth's troposphere. An array of magic-based superheroes use their mystical powers to prevent both the moon and the Earth from suffering geological catastrophes from their intense gravitational attraction. Our satellite now possesses a combustible atmosphere, and the evil Martians quickly surrender and submit to banishment to another dimension (the Phantom Zone, in fact) rather than being incinerated. Even if you grant all of the above as one major-league miracle exception, there is still a serious physics problem with this story line.

Newton's second law, $\mathbf{F} = \mathbf{ma}$, tells us that if a net force is applied to a mass, no matter how large, there will be a corresponding acceleration. By the late 1990s DC Comics had established that Superman was capable of lifting eight billion pounds. Let's assume that, given the enormous stakes, both Wonder Woman and Green Lantern exerted themselves to provide an equivalent force as they pulled on the moon. So the total force that these three heroes can supply is twenty-four billion pounds. Since the magic-based heroes are nullifying the effects of gravity, we'll assume that as the moon comes closer to the Earth there is no assist from Earth's gravitational field (this will keep the calculation at a simple level). The moon has a mass of nearly seventy billion trillion kilograms. Newton's law therefore indicates that the moon will indeed accelerate owing to this force, but the rate of change of motion will be very, very small. The acceleration of the moon will be 5 billionths feet/sec^2 (the acceleration due to gravity on the surface of the Earth is 32 feet/sec^2), and so it will take a very long time to displace the moon a significant distance. At this acceleration, the time needed for the moon to travel roughly 240,000 miles from its normal orbit to within our upper atmosphere is more than 735 years! We can only conclude that J'onn J'onzz performed some outstanding stalling in order to keep the evil Martians from realizing what was going on for more than seven centuries!

WITH THE WINGS OF AN ANGEL—COULD YOU FLY?

Another of the original members of the mutant team the X-Men introduced in 1963 was Warren Worthington III, whose mutant gift comprised two large, feathered wings growing out of his back. None of the other members of this superhero team possessed the power of flight, and aside from Ice Man's ice ramps, the Angel was the only character who could avoid walking or taking the bus when called upon to face off against the Brotherhood of Evil Mutants.[*] Other winged superheroes or villains, such as DC Comics's Hawkman or the Spider-Man villain the Vulture, used "antigravity" devices such as Hawkman's Nth metal, to overcome gravity. They employed their wings, that were connected to their backs in the case of Hawkman or Hawkgirl or sprouting from his arms for the Vulture, as

---

[*] And too few organizations nowadays, in my opinion, are proud enough of their "evil" status to boldly incorporate it into their title and stationery.

steering devices to help them maneuver while airborne. In contrast, the X-Men's Angel used his wings as his primary means of locomotion. It certainly seems reasonable that having wings growing out of your back would enable you to fly, but could it really?

Birds and planes manage to slip the surly bonds of gravity through the same physics principle: Newton's third law that for every action there is an equal and opposite reaction. A common misconception is that the pressure change induced by a fast-moving object (termed the Bernoulli effect), underlies how airplanes fly. We encountered this pressure differential when we considered the Flash dragging Toughy Boraz behind him in his super-speed wake in chapter 5. A fast-moving object such as the Scarlet Speedster must push the air out of his way as he runs, and consequently leaves a region of air with a lower density behind him. As the air races back to fill this partial vacuum, through the same principle as in our discussion of entropy in chapter 12, it will push anything in its way, such as the litter swirling behind fast-moving traffic or trains. However, if the difference in wind speed above and below the wing is a result of the wing's contour, then planes should not be able to fly upside down, because the pressure difference generated by the Bernoulli effect would tend to push the plane toward the ground.

In any case, we can always rely on Newton's third law, which tells us that forces always come in pairs. To provide an upward force on the airplane wing equal to or greater than the weight of the plane, an equivalent downward force from the wing must be applied to the air moving past it. The down draft of air in the region underneath the wing results in an upward lift that carries the plane into the wild blue yonder. When Superman leaps, he pushes down on the ground so that an equal and opposite force pushes back on him, starting him up and away. Similarly, birds flap their wings, pushing a quantity of air downward. The downward force of the wing on the air is matched by an upward force by the air on the wing. The greater the wingspan, the larger the volume of air displaced, and the greater the corresponding upward force. This is why it is impossible for Prince Namor the Sub-Mariner to fly using his tiny ankle wings. These petite wings are too small to provide sufficient lift to counter Namor's weight.

If Warren Worthington III weighs 150 pounds (equivalent to a mass of 68 kilograms), then his wings must provide a downward force on the air of at least 150 pounds, such that the air's reaction on his wings balances his weight and keeps him above the ground. Of course, if he wants to accelerate, then his wings have to provide a force greater than 150 pounds in order for there to be an excess force (upward lift minus downward weight due to gravity) to provide a net acceleration. If his wings provide an upward force of 200 pounds while gravity exerts a downward force of 150 pounds, then Warren experiences a net vertical force of 50 pounds. Force equals mass times acceleration, so this upward force of 50 pounds creates vertical acceleration of 11 feet/sec^2. With this acceleration the Angel will go from 0 to 60 mph in a little over eight seconds, neglecting the considerable air resistance that he would have to overcome. Once he stops flapping his wings, the only force acting on him is gravity pulling him back to Earth. Of course he can glide once airborne, but he must continue to apply a downward force on the air to truly fly and not coast.

Two hundred pounds is a considerable force for his wings to apply, but it is not unreasonable that a person could bench press 133 percent of his body weight. Birds such as the California condor or the wandering albatross weigh roughly thirty or twenty pounds respectively, and yet are able to generate sufficient force to fly. But Warren Worthington III is not built like a bird. Birds do not have wings growing out of their backs—their arms have evolved into wings. They have two additional modifications that assist their arm-wings: (1) They have a keeled sternum bone—that is, birds have a hinge built into the flat bone in the center of their chests that is comparable to your rib cage. This hinge acts as an anchor point for their other adaptation, namely (2) birds have two extremely large muscles, the supercorocoiderus and the pectoralis, used for beating their wings. Birds have so much breast meat because these largest muscles, their pectorals, provide the majority of the force to the wings in flight. Recall from chapters 8 and 10 that the strength of bone or muscle increases with its cross-sectional area. Consequently, the Angel must have enormous pectorals if he is to be able to use his wings to get off the ground. With a wingspan of 16 feet and a weight of 150 pounds, Warren has a weight-to-wingspan ratio of nine pounds per foot, in contrast to a ratio of three pounds per foot for a California condor. Warren's arms do not participate in supplying a force to his wings, and he must provide an upward lift using only his chest

and back muscles, making him a muscle-bound and fairly ineffective superhero.

There are other adaptations for flight that Warren could possess
that would require additional miracle exceptions. To reduce their
body weight, birds have lightweight bones, with a very porous
structure that yet remains remarkably strong. Birds also have very
efficient respiratory systems, so that every single oxygen molecule
residing in their lungs is replaced within two deep breaths. In contrast, with every breath we take, we only exchange 10 percent of the
air molecules residing within our lungs. Birds need to be able to rapidly refresh their air supply, as their breast muscles are working so
hard to maintain them aloft. Warren's breathing could be similarly
efficient. But for all this, unless he is to be drawn with enormous
pectoral muscles—more fitting for some of the female superhero
characters from the 1990s—the wings on his back are more orna-
mental than functional.

ENTER . . . THE VISION!

When Roy Thomas took over the writing duties of the Marvel
comic book *The Avengers* in the mid-1960s, he would frequently
reintroduce Golden Age characters with a new Silver Age twist,
just as DC Comics had done when they initiated the Silver Age.
One of the more popular characters created by Thomas and artist
John Buscema is the Vision. Originally a supernatural costumed
crime-fighter in the 1940s, the new Vision introduced in *Avengers*
57 is an android* created by Ultron, another android. Ultron is
one of the Avengers' most dangerous foes, and the Vision was initially intended to infiltrate the super team in order to destroy
them from within. Rebelling against his programming, the Vision
saved the lives of the Avengers and went on to become a valued
member of the team.

In addition to laser vision, the power of flight, and the mind of
a computer, the Vision possessed the superpower of total independent control of his body's density. He could make his body, or any
part thereof, as hard as diamond or so insubstantial that he could

---

* Technically the Vision is a "synthezoid," and no, I don't know what the
difference is.

pass through solid objects. Kitty Pryde of the X-Men walks through walls using her mutant ability to vary her quantum-mechanical tunneling probability, but the Vision should stick to using the door when he wants to enter a room.

The density of any object is defined as the mass per volume, and can be altered either by changing the mass or varying the volume. The volume is governed by the average spacing between atoms. Any solid typically has its atoms packed fairly closely, so the atoms can be considered to be touching (they have to be this close in order to form chemical bonds, which are what hold the atoms together in a solid after all). Very roughly, all solids have the same density, within a factor of ten or so. Diamond is a hard material not because the atoms are packed particularly closely, but because the chemical bonds holding the carbon atoms together are very rigid and inflexible. Graphite, used in pencil lead, has a chemical composition identical to diamond, but is very soft. Graphite's density is less than a factor of two lower than diamond's, but the big difference in its hardness arises from the weak chemical bonds holding the layers of hexagonal planes of atoms together.

Even if the Vision could control his density at will and could maintain the structural integrity of his body, he could not pass through walls. A gas, such as the air in your room, is comparatively dilute, with the average spacing between atoms being roughly ten times larger than the size of an atom. Yet the fact that the air in your room is less dense than the walls does not mean that the air can pass through the solid walls. Good thing, too, otherwise the air in an airplane would leak out through the fusilage and make air travel an even more unpleasant experience. We must therefore conclude that Ultron made a second error when he constructed the density-altering Vision (the first was believing that such a noble android would betray the mighty Avengers).

CAN THE ATOM USE THE TELEPHONE TO REACH OUT AND TOUCH SOMEONE?

The DC superhero the Atom has appeared throughout this book, and his ability to reduce his size and mass independently have provided excellent illustrations of a wide range of physical phenomena. Of course, occasionally his shrinking would take him to ridiculous extremes, such as whenever he visited other worlds that contained civilizations, cities, and advanced technology all residing within an atom. Given that there are nearly a trillion, trillion atoms in a cubic centimeter of a typical solid, it's amazing that the Atom ever managed to find these nanoworlds, unless they are a routine feature of every element in the periodic table. The implausibility of the Atom's powers was slyly acknowledged in 1989, in a scene in his second regular series, *The Power of the Atom* # 12. In this story the Atom shrinks both himself and a colleague in order to escape a supervillain's death trap, and they wind up decreasing to subatomic lengths in order to pass through the empty spaces in the floor's atoms. Pausing in their miniaturization, they sit on an electron, discussing the events of the past few issues. The Atom's friend notes that they are smaller than oxygen molecules and wonders, "How are we even breathing?" To which the Atom honestly replies, "I'm not sure."

Superman can fly, the Flash can run really fast, Hawkman has his wings and antigravity belt, Storm rides on thermally generated air currents, but how do you get around when you're very, very tiny? Ant-Man uses flying carpenter ants as his personal taxi service, the Wasp has wings that grow out of her back when she shrinks at constant density, but the Atom has Bell Telephone. There were two separate adventures in *Showcase* # 34, the comic that featured the debut of the Silver Age Atom. The first tale told the origin of the Atom, which we will address in the next section. In the second story, "Battle of the Tiny Titans," the Atom for the first time employs a unique mode of transportation. In this story he needs to confront a small-time crook named Carl Ballard who is clear across town. Presumably after looking up Ballard in the phone book, the Atom dials his number while setting up a metronome near the receiver, which creates a "tick-tock" sound. Shrinking himself smaller and smaller, the Mighty Mite jumps into one of the holes on the speaker of his telephone, and in the next panel we see him flying out of the receiver of Carl Ballard's phone.

The "explanation" for this trick is revealed on a text page in the back of the comic.* By dialing Ballard's phone number, the

---

* In the late 1950s and early 1960s, comics always included no fewer than two pages of prose in order to qualify for second-class postage mailing rates reserved for "magazines," which were defined as journals that contained at least two pages of text.

Atom causes an electrical impulse to travel from his phone to the central telephone exchange, which then forwards the signal to Ballard's phone. When the circuit is completed once Ballard answers the ringing phone, the signal—in this case the ticking metronome—is transmitted from the Atom's phone to Ballard's. At this point the Atom jumps into his speaker, shrinking down to the size of an electron, and rides these electrical impulses from his phone to Ballard's.

The writer of this text page, DC Comics editor Julie Schwartz, correctly describes how a telephone transfers sound into electrical impulses. A thin diaphragm vibrates when sound waves strike it, which in turn compress or dilate carbon granules that are adjacent to the membrane. The electrical conduction through the carbon grains is very sensitive to how tightly they press against one another. As you speak, the interconnections between the grains alternately contract or expand, and the electrical signal down the wire is appropriately modified. At the other end of the telephone connection, the electrical signal causes other carbon grains to undergo equivalent vibrations that are transferred to another diaphragm. The diaphragm's vibrations create pressure waves in the air that are then detected by the ear of the person receiving the call. All of this Julie Schwartz got right. Where he goofed is in assuming that the Atom could hitch a ride on the electrical impulses propagated down the wire.

When you speak, complex sound waves can convey all sorts of information. The sound waves can be detected by another membrane (such as an eardrum), causing it to vibrate in accordance with the amplitude, wavelength, and even phase information that is encoded in the message you spoke. But it is the *wave* that carries that information—not the air you expelled from your mouth. By speaking, you set up alternating regions of less dense and more dense air (equivalently you can think about the density variations as pressure modulations—a reasonable approximation at a constant temperature) that move away from the speaker. It is not the air coming from your mouth that reaches the listener; otherwise you would never have to worry about noisy neighbors in the apartment next door.

Similarly, the information encoded in electrical impulses in a telephone wire is transmitted by means of electromagnetic waves, rather than having the electrons move down the wire. What happens is that a region of higher than normal density electrons is unstable (as the negatively charged electrons repel each other) and expands into the adjacent regions, causing a buildup of electron density in the next spatial location, which in turn causes a bulge farther down the line, and so on. The speed of this transmission is determined by the electrostatic repulsion that pushes the electrons away from each other. That is, if I shake one electron, how long will it take a second electron some distance away to respond to the first electron's motion? Pretty quickly, as it turns out, as the electrical interaction between two charges in the wire is communicated at roughly one-third of the speed of light. Depending on the distance, there will be a barely perceptible time lag between moving the first charge and it being noticed by the second charge. The speed of light is so fast—186,000 miles per second—that this time lag will be less than a billionth of a second over a distance of 12 inches. If the Atom were riding on one electron carrying the electrical impulse signal along the telephone wire, he would have to jump to the next bunch of electrons with a response rate greater than the speed of light in order to "ride the wave" all the way to the receiver.

It's a good thing that the information in a telephone wire is in fact transmitted at the speed of light, as the average speed that an electron moves along a wire in response to an external electric field is less than a millimeter/sec, nearly a trillion times slower. If you had to wait for the electrons to physically travel along the telephone wires before your message could be sent, it would be quicker to just walk to the house of the person you're calling and speak to her face-to-face.

EVERY PHYSICIST'S SECRET SUPERPOWER

When not fighting crime as the Atom, Ray Palmer's civilian identity is equally heroic, for he is a physics professor at Ivy University. As mentioned in chapter 12, it was the late-night discovery of a strange meteorite that led to the research breakthrough that enabled Palmer to develop a second career as a costumed crime-fighter. As shown in fig. 39, Palmer discovers that the meteor is in fact a chunk of white-dwarf-star matter that will enable him to

Fig. 39. *Physics Professor Ray Palmer discovers the white-dwarf-star fragment that will turn out to be the key missing ingredient in his miniaturization device and eventually lead to his moonlighting as the superhero the Atom (from Showcase # 34).*

miniaturize himself and independently control his mass. Ray strains to lift and carry the meteorite, which is roughly twelve inches in diameter, over to his car. We are privy to Prof. Palmer's thoughts as he struggles with the great weight. "So heavy—I can hardly lift it! *Puff!* I don't know the odds against one white dwarf hitting another out in space—*Puff*—but it could happen—and when it did, this piece drifted until it landed in this field." (By the way, as also shown in fig. 39, in the mid-1960s physics professors typically drove Cadillac convertibles.)

Ray's reasoning here is sound. When a low mass star of a certain size has exhausted most of its elemental fuel, the energy released by fusion reactions is insufficient to counteract the gravitational pull of the star's core. The large force at the center of the star leads to a massive compression, until its density is three million grams per cubic centimeter, in which case we call the remnant a white dwarf. The pull of gravity on the remaining core of a white dwarf star is so great that only a cataclysmic explosion would generate sufficient energy to enable a small chunk of the core to break away from the rest of the star and float through space. If, as some astrophysicists have suggested, the light detected from a particular type of supernova explosion (labeled "Supernova Ia" events) results from the collision of two white-dwarf-star cores, then from the frequency of such supernovas we can say that white dwarf collisions occur roughly a dozen times a year.

As Ray reminds himself while struggling with the meteor fragment, the rock he is holding is heavy because it is composed of "degenerate" matter. The electrons are termed "degenerate" because they are all in one single quantum state, unlike in a normal star where the electrons would be distributed over many quantum states, corresponding to different energies. The interior of the white dwarf is composed of carbon and oxygen nuclei and a sea of electrons packed as closely as they can be. The core of white dwarfs cannot be easily compressed further, for all the electrons are already in the lowest possible energy state. This is what Ray means when, as he nears his car, he thinks to himself that white dwarf stars are composed of "degenerate matter from which the electrons have been stripped, greatly compressing them." The electrons are still in there, but are not associated with any particular atomic ions.

Ray is certainly correct that this "degeneracy" is why the white dwarf star is so dense. The rock Ray is carrying appears to have a radius of 6 inches. Assuming a spherical white dwarf fragment, the volume would be $(4\pi/3) \times (\text{radius})^3$. In this case the volume of the rock is $(4\pi/3) \times (6 \text{ inches})^3 = 905$ inches^3—equivalent to nearly 15,000 cm^3 since 1 inch equals 2.54 centimeters. To find the mass of the rock we multiply the density of white-dwarf-star matter (3 million grams/cm^3) by this volume (15,000 cm^3), which gives us 45 billion grams, which is equal to 45 million kilograms. Converting this mass to weight, we multiply the mass by the acceleration due to gravity ($W = mg$), and find that the meteorite in fig. 39 weighs one hundred million pounds. No wonder Prof. Palmer, physics professor at Ivy University, is huffing and puffing as he struggles with his find—that little rock weighs 50,000 tons!

But it turns out that this is, technically, *not* actually a blooper. Despite appearances, there is nothing wrong with the scene depicted in fig. 39. And that is because we physics professors are Just. That. Strong. Remember this the next time you're tempted to kick sand in someone's face at the beach. You never know if that seemingly ninety-eight-pound weakling actually has an advanced degree in physics.

AFTERWORD–

LO, THERE SHALL BE AN ENDING!

IT SHOULD COME AS NO SURPRISE that comic books and physics make a good match; after all, the fun underlying science is not so different from that of a good superhero comic-book story. In both situations either the scientist or the comic-book reader (in some cases they may be one and the same) are presented with a set of rules to be applied in novel, challenging situations. The rules may be Maxwell's equations of electricity and magnetism and Schrödinger's equation, and the challenging problem may be trying to develop a semiconductor analog of a vacuum tube. Alternatively, the rules may be that our hero can run at super-speed and has an aura that protects him from the adverse effects of air drag and electromagnetic induction, and the challenge would be that he has to capture a villain armed with a freeze gun capable of icing up any surface, while recovering the stolen bank funds and without harming any innocent bystanders. In both situations the trick is to find a solution that employs the known rules in a new way (if an old solution would work, we'd just use that), without utilizing anything that is deemed impossible under these guidelines. We can't design a transistor device that, in order to function, requires electrons to split into two halves or be attracted toward each other without an intervening positive charge, because the basic unit of negative charge has never been observed to behave in this fashion. Similarly, a Flash comic-book story featuring the Scarlet Speedster defeating Captain Cold by shooting heat beams from his eyes would be unsatisfying, as this is not an ability that the Flash has ever possessed.

The goal of basic scientific research is to elucidate the fundamental laws of nature, and the highest accomplishment is the discovery of a *new* rule or principle. Equally good is the clear demonstration of a violation in a preexisting rule, for new physics is discovered when we understand under what circumstances the old rules do not apply. Similarly there are times when an established comic-book character suddenly acquires a previously unsuspected ability, such as when Sue Storm of the Fantastic Four discovered in *Fantastic Four* # 22 that the cosmic ray bombardment that gave her the power of invisibility had also bestowed upon her the ability to generate "invisible force fields."* The dynamics between Sue and her teammates were radically altered following the discovery of this new superpower, and over the years she would learn to generate her force fields in an offensive as well as defensive capability.

But such cases are rare both in comic books and in real-world physics. There is, however, an unending stream of exciting and challenging problems in physics, just as there is an unlimited source of engaging comic-book stories waiting to be told. The two central ingredients are the same for both science and comic books: an understanding of the basic rules of the game and a fertile imagination.

Scientists don't typically consult comic books when selecting research topics (funding agencies tend to frown on grant proposals that contain too many citations to DC or Marvel comics) but the spirit of "What if . . ." or "What would happen when . . ." infuses both the best scientific research and comic-book adventures. To be sure, there are times when comic books and science fiction anticipate scientific discoveries, just as cutting edge research is occasionally employed as the springboard for superhero adventures (as in the aforementioned *JLA* # 19).

Sometimes it takes a while for the science to catch up with the comic books. As an example, consider the magician Abra Kadabra, a Flash villain who has plagued the Scarlet Speedster from nearly the beginning of his crime-fighting career. Garbed in the conventional

---

* This is a bit redundant, since every force field in comic books is invisible, save one. Given that Green Lantern's ring is powerless against anything colored yellow, the villain the Shark in *Green Lantern* # 24 had the ability to project "yellow invisible force fields"!

stage magician's attire of eveningwear and a top hat, he would use his "magic" to bedevil the Viceroy of Velocity, such as the time he turned him into a human marionette. However, it was revealed that Abra Kadabra was a scientist from the far future and that his "magic" in the twentieth century was actually sixty-fourth-century technology.* The creators of the Flash comics clearly subscribe to the notion that our present-day science and engineering would appear to be supernatural to those in the distant past. After all, imagine the reaction you'd receive if you could travel one thousand years into the past and display just a fraction of the appliances found in a modern home (assuming you also brought a power supply with you).

It was left deliberately vague in the Silver Age story as to how sixty-fourth-century science could transform someone into a living puppet. The "explanation" would have to wait until the late 1990s, where Kadabra informs us that he employed nanotechnology to restructure the Flash at the molecular level, demonstrating once again the trouble that a crooked ex-scientist can cause. Certainly nanometer-scale machines cannot cause such damage, but it's difficult to say what can and can't be done in another few thousand years, provided it doesn't involve a violation of established physics. As mentioned in chapter 7, a useful formula for anticipating scientific advances may be **(Science Fiction) + (Time) = Science**.

In fairness, however, the predictive ability of speculative fiction sometimes gets the technological aspects right, but widely misses other revolutions that have transformed our society. Consider for example the 1966 television program *Lost in Space*. This popular TV program envisioned a trip to the stars by the Robinson family, accompanied by an intelligent robot and Dr. Zachary Smith, a villainous and cowardly stowaway. The show first aired on September 15, 1965, and was imagined to take place in the distant future, all the way in October 1997. As pointed out in a *New York Times* article in 1997, discussing an anniversary rebroadcast

---

* Bored with the peaceful idyll that society had finally achieved, Kadabra traveled back to our time in order to wreak mischief. The desire to escape the monotony of future utopia motivated the Marvel Comics time-traveling villain Kang to visit our time, intent on world conquest. It appears that human nature—at least for some humans—will always rebel against a well-ordered, perfect society.

of the pilot episode, while the producers and writers of *Lost in Space* were not far wrong in assuming that thirty years hence starships and robots might be feasible, they goofed spectacularly regarding one very crucial aspect of modern life in the late 1990s.

A scene set in mission control as the starship is preparing to launch features a familiar bank of computer monitors manned by an array of nearly identical short-sleeved white-shirted engineers. At the elbow of each mission control engineer is a small metal disc that one would never, ever find in the NASA of today. The science-fiction writers in 1966 never imagined that in thirty years mission control would be a smoke-free environment and consequently no place for ashtrays. Thereby a cautionary note, that extrapolating potential scientific and technological innovations is duck soup compared to predicting future social customs.*

* * *

If the study of the natural world has demonstrated anything, it is that, unlike the Hulk, the smarter we get, the stronger we become. Now that you've finished this book, perhaps you'll feel a little stronger yourself, if not in arm, then at least in mind. Which is the only type of strength that really matters. It is our intelligence that provides the competitive advantage that enabled us to become the dominant species on the planet. We are not as fast as the cougar, cannot fly like the bird, and are not as strong as the bear or as indestructible as the cockroach. It is our intelligence that is our superpower, if you will. As quantum mechanics pioneer Niels Bohr said, "Knowledge is in itself the basis of civilization."

The optimism at the heart of all comic-book adventures lies within the scientific endeavor as well, as they both hold out the promise that we will overcome our physical challenges and improve the world. How science is to be employed, whether to ease hunger and cure disease, or to develop an army of killer robots, is up to us. For guidance in how to use our knowledge wisely and ethically, one could do worse than look to the stories in comic books. It is as true today as it was many years ago when Ben Parker

---

* Sometimes these changes take a while to come about. If ancient Egyptian galley slaves could visit a modern-day health club and see the rowing machines that wealthy (compared to themselves) free men and women employ, their heads would no doubt explode in shock!

said to his nephew Peter back in *Amazing Fantasy* # 15: "With great power there must also come—great responsibility." But responsibility to do what? One answer was provided by the Man of Tomorrow in the story "The Last Days of Superman" in *Superman* # 156. Believing that he was dying from an infection of Virus X (fortunately, a false alarm), Superman etched a farewell message to the people of Earth on the Moon with his heat-vision, a message he'd intended to be discovered after his demise. His final, parting words to the people of his adopted planet were: "Do good to others and every man can be a Superman."

Face front, true believer!

ASK DR. K*

In my class entitled The Physics of Superheroes, the following questions, not all of them physics-related, have come up repeatedly. While you may have differing opinions on some of these, as the professor, my answers are the correct ones.

WHO'S THE MOST REALISTIC SUPERHERO?

This is easy. Obviously it must be Batman, who always manages to find a way to win using just his razor-sharp mind and highly trained body. Though, given the number of times he has been knocked unconscious in his more than sixty-year crime-fighting career, he must have some hidden superpower that keeps him from developing permanent brain damage.

WHO'S THE MOST UNREALISTIC SUPERHERO?

This is also easy. Super-strength, super-speed, flight, invulnerability, super-hearing, X-ray vision, heat-vision, telescopic-vision, microscopic-vision, super-breath, super-ventriloquism, super-hypnotism, and he always obeys all the rules and has never tried to take over the world? Superman is totally unrealistic—and thank goodness for it!

---

* The "K" stands for Action!

WHAT'S THE PHYSICS BEHIND MAGIC-BASED SUPERHEROES?

There isn't any. In the Golden Age there were quite a few costumed heroes whose superpowers were magically based. Jerry
Siegel, who along with Joe Shuster introduced Superman to the
world, went on to co-create an even more powerful character:
The Spectre, an actual angel of vengeance. Suffice it to say that
there were few physics principles that *weren't* violated in a typical
Spectre story, what with the Spectre and his demon foes throwing
planets at each other and such. Other mystical heroes such as Dr.
Fate; Dr. Strange; Wonder Woman (an Amazon princess after all);
and the Norse god of thunder, the mighty Thor, all had what could
be more accurately described as "fantasy" rather than science-
fiction–based adventures, and with one notable exception, their
abilities and feats ignored scientific principles entirely.

MOST UNLIKELY YET PHYSICALLY ACCURATE SUPERHERO FEAT?

Surprisingly enough, a signature characteristic of a magic-based hero
turns out to be physically accurate, provided we allow the standard
miracle exception, of course. When the Norse god Thor needed to
travel quickly from one location to another, he used his great
strength to twirl his Uru hammer at high speed. Throwing it in the
direction he wanted to go, he would momentarily let go of the hammer's handle strap and then grab on to it again, flinging himself
through the air as an unguided missile. This would appear to be a
perfect violation of the principle of conservation of momentum. In
fact, in *Bartman Comics* # 3 (featuring the adventures of Bart Simp-
son's superhero alter-ego), Radioactive Man is so angered when he
spies a Thor-like character taking flight in this manner that he socks
the mock-Thor, intoning, "This is for breaking the laws of physics!"
And yet, such a means of transportation is physically plausible.

When Thor twirls his hammer, the mighty Mjolnir, he plants
his feet firmly on the ground, coupling his body's center of mass
to the Earth's. This is presumably what makes the X-Men villain

the Blob so difficult to move—his mutant ability enables him to strongly couple his center of mass to the Earth's, such that dislodging the Blob requires moving the entire Earth unless the connection is broken. When Thor is ready to let fly, all he must do is jump slightly (breaking his connection with the Earth) at the moment he throws his hammer in the desired direction. He doesn't even need to go through that business with releasing and regrabbing the handle strap. Inexperienced track athletes can confirm that losing one's footing during the hammer throw can result in an undesired short trip. If one is as strong as a thunder god, one can use this technique to fly through the air with the greatest of ease. No wonder they named a day of the week after this guy!

WHO IS FASTER: SUPERMAN OR THE FLASH?

The Flash.

WHAT IS ADAMANTIUM?

A defect-free covalently bonded metal.

The strength of materials is primarily governed by the nature of the chemical bonds holding the atoms together in the solid phase. The strongest chemical bonds are termed "covalent bonds," where the individual atoms quantum-mechanically share their outermost electrons with their atomic neighbors. In order to break these bonds, one must remove the electrons from all of the bonds connecting an atom to all of its neighbors—an energetically costly process. Normal metals do not have directional bonds holding them together, which is they are as easily bent as wires. Adamantium, the strongest material in the Marvel Comics universe, must somehow combine the electrical properties of normal metals with the strong covalent bonds found in diamonds. In addition, adamantium must be defect-free. Flaws or imperfections in a diamond occur at regions in the solid where atomic bonds, through strain or the inclusion of impurities, are broken or weakened, and it is at these locations where the covalent bonding network is easiest to break.

CAN WOLVERINE'S CLAWS CUT CAPTAIN AMERICA'S SHIELD?

No. Wolverine's claws are composed of adamantium, but Captain America's shield is a one-of-a-kind alloy of steel and vibranium. The latter is an extraterrestrial material brought to Earth when a meteorite crashed in the African nation of Wakanda, ruled by the superhero the Black Panther. Vibranium has the ability to absorb any and all sound and converts the energy carried in the sound wave into some other, not-well-specified, form. Sound waves are alternations in pressure or density, and in a solid, sound is transmitted through the vibrations of the atoms. Vibranium possibly converts the atomic vibrations from an absorbed sound wave into an optical transition (although in the infrared portion of the spectrum, since vibranium does not seem to glow when used), thereby conserving energy in the process. The material that forms Captain America's shield was the result of a laboratory fluke, when a steel alloy and vibranium were accidentally fused into an alloy. The conditions under which this metallurgical fusion occurred were not recorded, and this synthesis has never been repeated. Vibranium's ability to absorb vibrations, coupled with the steel alloy's rigidity, is no doubt why all those who deal with Cap's mighty shield must yield.

CAN YOU TRAVEL THROUGH TIME BY RUNNING FASTER THAN THE SPEED OF LIGHT?

No. Theoretical physicists have hypothesized the existence of particles termed "tachyons" that can never travel *slower* than the speed of light, for which the direction of time would seem to be reversed. Tachyons were proposed as a test of certain consequences of the Special Theory of Relativity, such as violations in causality. As far as we know, they do not exist, and more importantly, even if they were as common as crabgrass it doesn't appear they can interact with our physical world, in which no object can move *faster* than the speed of light. The Flash may travel backward and forward in time using his Cosmic Treadmill, but its only real value is in providing the Scarlet Speedster with a cardio workout.

HOW IS PRINCE NAMOR'S TITLE PRONOUNCED?

It's Sub-Mariner, not Submarine–er. Prince Namor's creator Bill Everett joined the merchant marines at age fifteen (and left two years after) and presumably was very familiar with this synonym for seamen. His half-human/half-Atlantean hero, capable of breathing underwater, is best described as a *sub*-merged *mariner*. Imperius Rex!

WHAT'S THE DEAL WITH THE HULK'S PANTS?

When nuclear physicist Robert Bruce Banner was belted with gamma rays, he gained the ability to transform into an eight-foot-tall, 2,000-pound jade giant. As Banner undergoes his metamorphosis, his shirt, shoes, socks, and all other apparel are ripped to shreds—except for his stylin' purple pants. In Marvel comics it is suggested that Banner's slacks are composed of unstable molecules, invented by Reed Richards for the Fantastic Four's jumpsuits. This miracle fabric expands or contracts as does its wearer. Chemists will tell you that "unstable molecules" do indeed exist— they are the ones that fall apart because they are unstable. But the truth is that the Hulk's pants stay on thanks to an agency more powerful than gamma radiation—the Comics Code Authority.

RECOMMENDED READING

INTRODUCTION

There are many excellent reviews of the early history of comic books. In addition to the books explicitly cited in the text and listed below, I would recommend: *Men of Tomorrow: Geeks, Gangsters, and the Birth of the Comic Book* by Gerard Jones (Basic Books, 2004); *Tales to Astonish: Jack Kirby, Stan Lee, and the American Comic Book Revolution* by Ronin Ro (Bloomsbury, 2004); and *Great American Comic Books* by Ron Goulart (Publications International, 2001). Jim Steranko's excellent two-volume *The Steranko History of Comics* (Supergraphics, 1970, 1972) is worth searching out for his thorough and entertaining elucidation of the lineage from pulp heroes to comic-book superheroes. Les Daniels has written extensively and elegantly on the history of comic-book characters and his *DC Comics: Sixty Years of the World's Favorite Comic Book Heroes* (Bulfinch Press, 1995); *Superman: The Complete History* (Chronicle Books, 1998); *Batman: The Complete History* (Chronicle Books, 2004); *Wonder Woman: The Complete History* (Chronicle Books, 2001); and *Marvel: Five Fabulous Decades of the World's Greatest Comics* (Harry N. Abrams, 1991) are all highly recommended, as is *Silver Age: The Second Generation of Comic Book Artists* by Daniel Herman (Hermes Press, 2004). A historical analysis of the role of comic books in American popular culture is presented in *Comic Book Nation* by Bradford W. Wright (Johns Hopkins University Press, 2001).

While not explicitly a history of comic books, *Baby Boomer Comics: The Wild, Wacky, Wonderful Comic Books of the 1960s* by Craig Shutt (Krause Publications, 2003) is a fun overview of some of the high and low points of Silver Age comic books.

Others have explored the science underlying comic-book superheroes, and any reader disappointed that their favorite character was not sufficiently discussed here may try consulting *The Science of the X-Men* by Linc Yaco and Karen Haber (ibooks, 2000); *The Science of Superman* by Mark Wolverton (ibooks, 2002); *The Science of Superheroes* by Lois Gresh and Robert Weinberg (Wiley, 2002); and *The Science of Supervillains* by the same authors and publisher (2004). The science underlying other pop-cultural subjects has been explored in *The Physics of Star Trek*, by Lawrence Krauss (Basic Books, 1995); *The Science of Star Wars* (St. Martin's Press, 1998) and *The Science of the X-Files* (Berkley, 1998) both by Jeanne Cavelos; *The Physics of Christmas* by Roger Highfield (Little, Brown & Company, 1998); as well as his *The Science of Harry Potter* (Viking, 2002).

Those readers interested in a deeper discussion of the philosophy and nature of physics investigations should consider Richard Feynman's *The Character of Physical Law* (Random House, 1994) and *The Pleasure of Finding Things Out: The Best Short Works of Richard P. Feynman* (Perseus Publishing, 2000); as well as Milton A. Rothman's *Discovering the Natural Laws: The Experimental Basis of Physics* (Dover, 1989) and *The Fermi Solution: Essays on Science* by Hans Christian von Baeyer (Dover 2001).

SECTION ONE—MECHANICS

While this book covers many of the topics treated in an introductory physics class, those readers who are gluttons for punishment and wish to consult a traditional physics textbook (or seek to verify that I am not trying to pull any fast ones) may find *Conceptual Physics* by Paul G. Hewitt (Prentice Hall, 2002) helpful. It is written as a high school physics text, so the mathematics remains at the algebra level. An abridged version of Richard Feynman's brilliant lectures in physics, covering the basis of classical physics, *Six Easy Pieces* (Perseus Books, 1994) is highly recommended.

There are several excellent biographies of Isaac Newton. The reader interested in learning more about this towering intellect may consider *The Life of Isaac Newton* by Richard Westfall (Cambridge University Press, 1994); *Newton's Gift* by David Berlinski (Touchstone, 2000); and *Isaac Newton* by James Gleick (Pantheon Books, 2003).

The discussion of the Special Theory of Relativity in chapter 6 went by so fast that its brevity can be attributed to Lorentz contraction. The *first* book anyone interested in this subject should read is *What Is Relativity* by L. D. Landau and E. B. Romer (translated by N. Kemmer) (Dover, 2003), which in only 65 pages (with figures!) clearly explains, without equations, the physical concepts underlying Einstein's theory. Fuller discussions of this fascinating subject can be found in: *Relativity and Common Sense* by Hermann Bondi (Dover Publications, 1962); *An Introduction to the Special Theory of Relativity* by Robert Katz (D. Van Nostrand Co., 1964); *Introduction to Special Relativity* by James H. Smith (W. A. Benjamin, 1965); and *Discovering the Natural Laws: The Experimental Basis of Physics* by Milton A. Rothman (Dover, 1989). Be warned that all of these treatments deal with the mathematics underlying relativity as well as the physical concepts.

SECTION TWO—ENERGY—HEAT AND LIGHT

Excellent overwiews for the nonspecialist on how energy is created and transformed, particularly at the molecular level, can be found in *The Stuff of Life* by Eric P. Widmaier (W. H. Freeman & Company, 2002); *The Machinery of Life* by David S. Goodsell (Springer-Verlag, 1992); and *Stories of the Invisible* by Philip Ball (Oxford University Press, 2001). Background information on this mysterious quantity are available in *Energies: An Illustrated Guide to the Biosphere and Civilization* by Vaclav Smil (MIT Press, 1998) and *Energy: Its Use and the Environment* by Roger A. Hinrichs and Merlin Kleinbach (Brooks Cole, 2001), *Third Edition*. This last is a textbook written at a practically math-free level, with abundant information concerning the environmental issues involved in energy transformation.

Excellent popular accounts of the fascinating history of thermodynamics are: *A Matter of Degrees* by Gino Segre (Viking, 2002); *Understanding Thermodynamics* by H. C. Van Ness (Dover Publications, 1969); and *Warmth Disperses and Time Passes: The History of Heat* by Hans Christian von Baeyer (Modern Library, 1998). Issues related to the measurement of temperature are considered in an accessible manner in *Temperatures Very Low and Very High* by Mark W. Zemansky (Dover Books, 1964), while phase transitions are discussed in *The Periodic Kingdom* by P. W. Atkins (Basic Books, 1995); and *Gases, Liquids and Solids* by D. Tabor (Cambridge University Press, 1979).

Popular accounts of the history of electricity and magnetism are found in: *Electric Universe: The Shocking True Story of Electricity* by David Bodanis (Crown, 2005); *The Man Who Changed Everything: The Life of James Clerk Maxwell* by Basil Mahon (John Wiley & Sons, 2003); and *A Life of Discovery: Michael Faraday, Giant of the Scientific Revolution* by James Hamilton (Random House, 2002).

SECTION THREE—MODERN PHYSICS

There are many excellent overviews of quantum physics written for the nonspecialist. Highly recommended are: *Thirty Years That Shook Physics: The Story of Quantum Theory* by George Gamow (Dover Press, 1985) and *The New World of Mr. Tompkins* by G. Gamow and R. Stannard (Cambridge University Press, 1999).

Excellent, clear discussions of cutting-edge research in string theory can be found in: *The Elegant Universe* by Brian Greene (W. W. Norton, 1999); *The Fabric of the Cosmos* by Brian Greene (Alfred A. Knopf, 2003); *The Future of Spacetime* by Stephen W. Hawking, Kip S. Thorne, Igor Novikov, Timothy Ferris, and Alan Lightman (W. W. Norton and Company, 2002), and *Quintessence: The Mystery of Missing Mass in the Universe* by Lawrence Krauss (Basic Books, 2000).

The solid-state physics revolution that has transformed all of our lives is documented in the highly readable *Crystal Fire: Birth of the Information Age* by Michael Riordan (Norton, 1997) and *The Chip: How Two Americans Invented the Microchip and Launched a Revolution* by T. R. Reid (Simon & Schuster, 1985).

SUMMARY

In the spirit of continuing a review of the topics addressed here, the reader should consider these fun books employing the question-and-answer approach to cover a wide range of physics for the non-expert: *The Flying Circus of Physics with Answers* by Jearl Walker (Wiley, 1977) and *Mad About Physics: Braintwisters, Paradoxes, and Curiosities* by Christopher P. Jargodzski and Franklin Potter (John Wiley & Sons, 2000). In a similar vein, for those no longer intimidated by mathematics is *Back-of-the-Envelope Physics* by Clifford Swartz (Johns Hopkins University Press, 2003). Those readers eager to put their physics knowledge to use are directed to *How Does it Work?* by Richard M. Koff (Signet, 1961) and Cy Tymony's *Sneaky Uses for Everyday Things* (Andrews McMeel Publishing, 2003), which contains instructions for making your own Power Ring!

Finally, some comic-book recommendations. Both DC and Marvel have comprehensive reprint lines, where comics from the Golden Age through the present are collected, frequently on better-quality paper than the originals and at a fraction of their cost if you were to buy the back-issues separately today. The *Archives* series from DC and the *Marvel Masterworks* volumes reprint Golden and Silver Age comics focusing on a given character or team in a hard-cover format. In addition, Marvel has a line of paperback reprints termed *Essentials*, where twenty or so issues of Silver Age or later comics featuring a given character or title are reprinted on cheaper paper, in black and white, at a cost of less than a dollar per issue. Those readers whose memories of former favorites have been jogged or those who have developed a new interest will almost certainly find a reprint volume at either your favorite bookstore or your friendly neighborhood comic-book shop. To find the nearest comic-book store, dial 1-888-COMICBOOK or visit http://csls.diamondcomics.com.

There are some collections, however, that should be considered required reading as part of any well-rounded liberal education in costumed superheroes. At the top of the list would be *Watchmen* (DC comics, 1986, 1987) by Alan Moore and Dave Gibbons, which is justifiably characterized as the *War and Peace* of comic books by the film director Terry Gilliam. For legal reasons the characters

in this story are disguised versions of Silver Age heroes originally
published by Charlton Comics (such as the Question, Blue Beetle,
Captain Atom, etc.) and a familiarity with them is not necessary to
enjoy the story. These characters' adventures are now published by
DC Comics, where they are undisturbed by the fate that their dou-
bles met in Moore's and Gibbons's epic. Another must-read is
Frank Miller's *The Dark Knight Returns* (DC Comics, 1997),
which imagines a possible future fate for Batman. This miniseries
is considered by most to be responsible for saving Batman from
cancellation or worse—irrelevance—by returning the character to
his darker, grim and gritty roots and has set the tone for various
motion-picture versions of the Caped Crusader. Continuing the
concept of possible futures of superheroes, Mark Waid's and Alex
Ross's *Kingdom Come* miniseries (DC Comics, 1998) investigates
the interactions between DC Comics's superpowered heroes and
villains and normal civilians. The influence of Marvel Comics superheroes on society, seen from the point of view of a non-
superpowered photographer for the *Daily Bugle*, is explored in
Marvels (Marvel Comics, 2004) by Kurt Busiek and Alex Ross.
One of the best time-travel adventures can be found in the collection *Days of Future Past* (Marvel Comics, 2004) starring many of
the characters from the popular X-Men films, where Kitty Pryde
goes back in time to prevent a political assassination that will lead
humanity to a dark, dystopian future. Finally, to cleanse the palate
of all these deconstructions of the superhero myth, read Darwyn
Cooke's *DC: The New Frontier* Vols. 1 and 2 (DC Comics, 2004,
2005)—a brilliant *re*construction of the dawn of the Silver Age set
in the Cold-War America of the late 1950s when these heroes first
appeared.

KEY EQUATIONS

NEWTON'S THREE LAWS OF MOTION
Page 25

The basic principles of dynamics, as elucidated by Sir Isaac Newton, state that (1) an object at rest will remain at rest, or if in uniform straight-line motion, will remain in motion, unless acted upon by an external force; (2) if an external force does act on the object, then its change in motion (either speed or direction) is proportional to the outside force, that is $\mathbf{F} = \mathbf{ma}$; and (3) forces always come in pairs, commonly expressed as for every action there is an equal and opposite reaction.

DEFINITION OF ACCELERATION
Page 25

Acceleration is defined as the rate of change of velocity—either its magnitude (speed) or its direction, and has units of (distance/time)/time or distance/(time)^2.

WEIGHT = Mg
Page 26

A consequence of Newton's second law $(\mathbf{F} = \mathbf{ma})$ when the external force is the gravitational attraction of a planet. The force is then referred to as **Weight**, and the acceleration due to gravity is relabeled by the letter "**g**."

$$V^2 = 2gH$$
Page 30

A description of the velocity **v** of an object moving under the influence of gravity, whether slowing down as it rises or speeding up as it falls, through a distance **h**.

NEWTON'S LAW OF GRAVITATIONAL ATTRACTION
Page 33

The simple expression, also elucidated by Sir Isaac Newton, for the attractive force between any two massive objects. The force is proportional to the product of each object's mass and inversely proportional to the square of the distance that separates them.

$$g = GM/R^2$$
Page 36

A consequence of Newton's law of gravitational attraction is that the acceleration due to gravity for any large object such as a planet or moon can be expressed as a universal constant **G** (**G** = 66.7 trillionth of $m^3/kg\text{-}sec^2$) multiplied by the object's mass **M**, divided by the square of the object's radius. This expression in only correct for spherically symmetric masses.

$$g_K/g_E = \rho_K R_K/\rho_E R_E$$
Page 36

By making use of the fact that the mass **M** of a planet can be written as the product of its density **ρ** and its volume ($4\pi R^3/3$ for a sphere), the acceleration due to gravity **g** = **GM/R^2** simplifies to $[\frac{4\pi}{3}]$ **GρR**, and when taking the ratio of the accelerations due to gravity for two planets, the constants **G(4π)/3** cancel out.

FORCE X TIME = CHANGE IN MOMENTUM
Page 49

A restatement of Newton's second law (**F** = **ma**) where the acceleration is the change in velocity divided by the time over which the external force acts. The momentum is defined as the product of the mass and velocity of an object.

CENTRIPETAL ACCELERATION $a = V^2/R$
Page 55

An object moving with velocity **v** in a circular arc with radius **R** is characterized by an acceleration for its continually changing direction. The magnitude of this acceleration is v^2/R and an external force pointing in toward the center of the circular trajectory of magnitude $F = mv^2/R$ must act on the object to account for this changing motion.

WORK = FORCE X DISTANCE
Page 118

Work in physics is another expression for energy, and any change in the kinetic energy of an object must result from an external force acting over a given distance. The expression indicates that you do no work when you hold a weight over your head, for while you do supply a force, there is no displacement of the stationary object. This is in conflict with the common usage of the term "work" but is correct physically—once you have increased the object's potential energy by raising it a distance over your head, there is no additional change in its energy if you maintain it in this raised state indefinitely.

KINETIC ENERGY = $(1/2)MV^2$; POTENTIAL ENERGY = MgH
Page 119

Expressions for the energy associated with motion (**kinetic energy** = $(1/2)mv^2$) or for the potential of motion in a gravitational field (**potential energy** = mgh). Note that the expression for potential energy is the same as the Work done raising an object of weight **mg** a height **h**.

FIRST LAW OF THERMODYNAMICS
Page 135

Essentially a restatement of the principle of conservation of energy, indicating that any change in the internal energy of a system will be the result of any Work done on or by the system, and any heat flow into or out of the system.

SECOND LAW OF THERMODYNAMICS
Page 140

In any process to convert the heat energy that flows from a hot object to a colder object into Work (defined as the product of force times distance), there will inevitably be some loss. That is, one cannot transform 100 percent of the heat flow into productive work. This is related to the entropy of the systems involved, which is a measure of the disorder of their components.

THIRD LAW OF THERMODYNAMICS
Page 142

Upon lowering the temperature of a system in equilibrium, which is a measure of the average energy of its components, the disorder of the system also decreases. The entropy of any system is only zero if there is only one configuration that it can have, and that state is only realized when the average energy of each component is zero—that is, at a temperature of Absolute zero.

COULOMB'S LAW OF ELECTROSTATIC ATTRACTION
Page 170

The mathematical expression for the force between two charged objects, indicating that the force is proportional to the product of each object's charge and divided by the square of the distance separating them. The formula is algebraically identical to Newton's expression for gravitational force. However, while gravity is always attractive, the force between two charged objects can be attractive if they have opposite signs (positive and negative) or repulsive if they have the same sign (both positive or both negative).

OHM'S LAW V = IR
Page 203

Expression relating the voltage **V** pushing or pulling electrical charges in a conductor of resistance **R** to the current **I** (number of charges moving past a given point per unit time). While this expression holds for most metals, not every electronic device obeys this simple linear relationship.

ENERGY = hf
Page 220

The quantum hypothesis that states that the change in energy of any atomic system characterized by a frequency **f** can only occur in steps of magnitude **Energy = hf,** where **h** is Planck's constant, a fundamental constant of nature. When a system lowers or raises its energy by emitting or absorbing light, it must do so through quantized packets of energy termed "photons."

DEBROGLIE RELATIONSHIP Pλ = h
Page 222

The motion of any matter having a momentum **p** is associated with a matter-wave of wavelength **λ**, where the product of the momentum and the wave's wavelength is Planck's constant **h**.

SCHRÖDINGER EQUATION
Page 238

The fundamental wave equation for the motion of quantum objects. By knowing the potential **V** acting on the object, one can solve this equation to obtain the wavefunction **ψ** that characterizes its behavior. Squaring this wavefunction yields the probability density of finding the object at a given point in space and time, and from this probability density, the average or expected values for any measurable quantity (location, momentum, etc.) can be obtained.

NOTES

INTRODUCTION

Page 2 *Action* # 333 (National Comics. 1966), uncredited.

Page 4 *World's Finest* # 93 (National Comics, 1958), reprinted in World's Finest Comics Archives Volume 2 (DC Comics, 2001). Written by unknown, drawn by Dick Sprang.

Page 6 "middle class sensibilities" *Comics, Comix & Graphic Novels: A History of Comics,* Roger Sabin (Phaidon Press, 1996).

Page 7 "Yellow journalism" *The Classic Era of American Comics,* Nicky Wright (Contemporary Books, 2000).

Page 7 "firmly established until 1933" *Comic Book Culture: An Illustrated History,* Ron Goulart (Collectors Press Inc., 2000).

Page 7 "big money in the Depression" *The Pulps: Fifty Years of American Pop Culture,* compiled and edited by Tony Goodstone (Chelsea House, 1970).

Page 8 "Superman was brain child" *The Illustrated History of Superhero Comics of the Golden Age,* Mike Benton (Taylor Publishing Co., 1992); *Superman. The Complete History,* Les Daniels (Chronicle Books, 1998).

Page 11 "before someone noticed and complained" *Men of Tomorrow: Geeks, Gangsters and the Birth of the Comic Book,* Gerard Jones (Basic Books, 2004).

Page 11 "Dr. Fredric Wertham's 1953 . . ." *Seduction of the Innocent,* Fredric Wertham (Rinehart Press, 1953).

Page 11 "The U.S. Senate Subcommittee" *Seal of Approval, The History of the Comics Code,* Amy Kiste Nyberg (University of Mississippi Press, 1998).

Page 12 "Declining sales from the loss" *Comic Book Nation,* Bradford W. Wright (Johns Hopkins University Press, 2001).

Page 13 *The Atom* # 21 (National Comics, Oct./Nov. 1965). Written by Gardner Fox, drawn by Gil Kane.

Page 13 "Give us back our eleven days!" *Encyclopedia Britannica* (William Benton, Chicago) vol. 4, pg. 619 (1968).

Page 13 *Brave and the Bold* # 28 (National Comics, 1960), reprinted in Justice League of America Archives Volume 1 (DC Comics, 1992). Written by Gardner Fox, drawn by Mike Sekowsky.

Page 14 "Why take the time . . . ?" *Man of Two Worlds, My Life in Science Fiction and Comics*, Julius Schwartz with Brian M. Thomsen (Harper-Entertainment, 2000).

Page 14 "The Hugo Award winner Alfred Bester . . ." *Star Light, Star Bright*, Alfred Bester (Berkley Publishing Company, 1976).

Page 14 "as reflected in this joke:" Lance Smith, private communication (2001).

Page 16 "physics is not about having memorized . . ." Hellmut Fritszche, private communication (1979).

CHAPTER I

Page 21 *Superman* # 1 (National Comics, June 1939), reprinted in Superman Archives Volume 1 (DC Comics, 1989). Written by Jerry Siegel and drawn by Joe Shuster.

Page 22 *Superman* # 330 (DC Comics, Dec. 1978). Written by Martin Pasko and Al Shroeder and drawn by Curt Swan and Frank Chiaramonte.

Page 22 *Action* # 262 (National Comics, 1960). Written by Robert Bernstein and drawn by Wayne Boring.

Page 23 "In his very first story . . ." *Action* # 1 (National Comics, June 1938), reprinted in *Superman* # 1 (National Comics, June 1939), reprinted in Superman Archives Volume 1 (DC Comics, 1989). Written by Jerry Siegel and drawn by Joe Shuster.

Page 23, 24 "By the 1950's" FN "How a radio-active element" *Superman: The Complete History*, Les Daniels (Chronicle Books, 1998).

Page 24 "Whether we describe the trajectory . . ." *The Principia: Mathematical Principles of Natural Philosophy*, Sir Isaac Newton, translated by I. Bernard Cohen and Anne Whitman (University of California Press, 1999); *Newton's Principia for the Common Reader*, S. Chandrasekhar (Oxford University Press, 1995).

Page 31 *Action* # 23 (National Comics, 1940), reprinted in Superman: The Action Comics Archives Volume 2 (DC Comics, 1998). Written by Jerry Siegel and drawn by Joe Shuster and the Superman Studio.

CHAPTER 2

Page 33 "As if describing the laws of motion . . ." *The Principia: Mathematical Principles of Natural Philosophy*, Sir Isaac Newton, translated by

I. Bernard Cohen and Anne Whitman (University of California Press, 1999);
Newton's Principia for the Common Reader, S. Chandrasekhar (Oxford
University Press, 1995).

Page 34 "This is the true meaning . . ." *The Life of Isaac Newton*,
Richard Westfall (Cambridge University Press, 1994); *Newton's Gift*, David
Berlinski (Touchstone, 2000); *Isaac Newton*, James Gleick (Pantheon
Books, 2003).

Page 36 "cubical planets such as the home world of Bizarro" *Superman:
Tales of the Bizarro World* trade paperback (DC Comics, 2000).

Page 38 "While planets in our own solar system" *Astronomy. The
Solar System and Beyond* (2nd edition), Michael A. Seeds (Brooks/Cole,
2001).

Page 39 "To be precise 73 percent of the" *Just Six Numbers. The Deep
Forces that Shape the Universe*, Martin Rees (Basic Books, 2000).

Page 40 "The fusion process speeds up as the star generates . . ." The
time necessary for iron and nickel synthesis can vary from several weeks to
less than a day, depending on the star's mass. See "The Evolution and Explosion of Massive Stars," S. E. Woolsey and A. Heger, *Rev. Modern Physics*
74, p. 1015 (Oct. 2002).

Page 41 "Only eight years earlier . . ." *Chandra: A Biography of S. Chandrasekhar*, Kameshwar C. Wali (University of Chicago Press, 1991).

CHAPTER 3

Page 43 "This all changed with a golf game" *Man of Two Worlds, My
Life in Science Fiction and Comics*, Julius Schwartz with Brian M. Thomsen (HarperEntertainment, 2000).

Page 43 "Instead he created a new superhero team from whole cloth."
Stan Lee and the Rise and Fall of the American Comic Book, Jordan
Raphael and Tom Spurgeon (Chicago Review Press, 2003); *Tales to Aston-
ish: Jack Kirby, Stan Lee and the American Comic Book Revolution* by
Ronin Ro (Bloomsbury, 2004).

Page 43 "Footnote" *Alter Ego* # 26, pg. 21 (TwoMorrows Publishing, July
2003).

Page 43 *Fantastic Four* # 1 (Marvel Comics, 1961), reprinted in Marvel
Masterworks: Fantastic Four Volume 1 (Marvel Comics, 2003). Written by
Stan Lee and Jack Kirby.

Page 43 *Amazing Fantasy* # 15 (Marvel Comics, 1962), reprinted in Marvel Masterworks: Amazing Spider-Man Volume 1 (Marvel Comics, 2003).
Written by Stan Lee and drawn by Steve Ditko.

Page 45 *Amazing Spider-Man* # 42–44 (Marvel Comics, 1964), reprinted
in Marvel Masterworks: Amazing Spider-Man Volume 5 (Marvel, 2004).
Written by Stan Lee and drawn by Steve Ditko.

Page 45 *Amazing Spider-Man* # 121 (Marvel Comics, June 1973), reprinted in *Spider-Man: The Death of Gwen Stacy* trade paperback (Marvel Comics, 1999). Written by Gerry Conway and drawn by Gil Kane.

Page 45 *Amazing Spider-Man* # 39 (Marvel Comics, Aug. 1964), reprinted in Marvel Masterworks: Amazing Spider-Man Volume 4 (Marvel Comics, 2004). Written by Stan Lee and drawn by John Romita.

Page 47 "This question was listed . . ." *Wizard: The Comics Magazine* # 100 (Gareb Shamus Enterprises, Jan. 2000).

Page 48 The towers of the George Washington Bridge are actually 604 feet above the water. See *The Bridges of New York*, Sharon Reier (Dover, 2000).

Page 50 "Col. John Stapp rode an experimental" *Wings & Airpower* magazine, Nick T. Spark (Republic Press, July 2003).

Page 51 *Spider-Man Unlimited* # 2 (Marvel Comics, May 2004). Written by Adam Higgs and drawn by Rick Mays.

Page 51 *Wizard: The Comics Magazine* # 104 (Gareb Shamus Enterprises, Apr. 2001).

Page 51 *Peter Parker: Spider-Man* # 45 (Marvel Comics, Aug. 2002). Written by Paul Jenkins and drawn by Humberto Ramos.

CHAPTER 4

Page 55 "Dragline silk webbing . . ." "Stronger than Spider Silk," Eric J. Lerner, *The Industrial Physicist*, vol. 9, no. 5, p. 21 (Oct./Nov. 2003); *Nature*, vol. 423, pg. 703 (2003).

Page 55 "Spider-Man is able to alter the material properties . . ." *Spider-Man Annual* # 1 (Marvel Comics, June 1963); reprinted in Marvel Masterworks: Amazing Spider-Man Vol. 2 (Marvel Comics, 2002). Written by Stan Lee and drawn by Steve Ditko.

Page 55 "Similarly, real spiders can control" C. L. Craig et al. *Molecular Biol. Evolution*, Vol. 17, 1904 (2000); Frasier I. Bell, Iain J. McEwen and Christopher Viney, A. B. Dalton, S. Collins, E. Munoz, J. M. Razal, V. H. Ebron, J. P. Ferraris, J. N. Coleman, B. G. Kim and R. H. Baughman, *Nature*, vol. 416, p. 37 (2002).

Page 55 "recent genetic engineering experiments . . ." *The Goat Farmer Magazine*, May 2002 (Capricorn Publications, New Zealand).

Page 56 "other scientists have reported . . ." D. Huemmerich, T. Scheibel, F. Vollrath, S. Cohen, U. Gat, and S. Ittah, *Current Biology*, vol. 14, no. 22, p. 2070 (Nov. 2004).

Page 56 "The silk-producing gene . . ." Jurgen Scheller, Karl-Heinz Guhrs, Frank Grosse, and Udo Conrad, *Nature Biotechnology*, vol. 19, no. 6, p. 573 (June 2001); S. R. Fahnestock and S. L. Irwin, *Appl. Microbiol. Biotechnol.* Vol. 47, p. 23 (1997).

Page 56 "As Jim Robbins discussed . . ." "Second Nature," Jim Robbins, *Smithsonian*, vol. 33, no. 4, p. 78 (July 2002).

CHAPTER 5

Page 57 "It was a dark and stormy night . . ." *Showcase* # 4 (National Comics, Oct. 1956), reprinted in Flash Archives Vol. One (DC Comics, 1996). Written by Robert Kanigher and drawn by Carmine Infantino.

Page 57 Footnote *Flash* # 110 (National Comics, Dec.–Jan. 1960), reprinted in Flash Archives volume 2 (DC Comics, 2000). Written by John Broome and drawn by Carmine Infantino.

Page 59 "Captain Cold, one of the first . . ." See, for example, *Showcase* # 8 (National Comics, June 1957), reprinted in Flash Archives Vol. One (DC Comics, 1996). Written by John Broome and drawn by Carmine Infantino.

Page 59 "While friction's basic properties were . . ." *History of Tribology*, 2nd edition, Duncan Dowson (American Society of Mechanical, 1999).

Page 62 "One can, of course, move faster than the speed of sound . . ." "Breaking the Sound Barrier," Chuck Yeager, *Popular Mechanics* (Nov. 1987); *Yeager: An Autobiography*, Chuck Yeager (Bantam, reissue edition, 1986).

Page 64 "This mechanism was recently proposed . . ." D. Hu, B. Chan, and J. W. M. Bush, *Nature* 424, pp. 663–666 (2003).

Page 64 *Flash* # 117 (National Comics, Dec. 1960), reprinted in Flash Archives volume 3 (DC Comics, 2002). Written by John Broome and drawn by Carmine Infantino.

Page 65 Fig. 10 *Flash* # 117 (second story) (National Comics, Dec. 1960), reprinted in Flash Archives volume 3 (DC Comics, 2002). Written by Gardner Fox and drawn by Carmine Infantino.

Page 66 "Were you to wrap your car . . ." The Discovery Channel program *Mythbusters* has recently debunked this urban legend, raising hopes that the long arm of the law may finally catch up to Spud Man.

Page 67 *Flash* # 124 (National Comics, Nov. 1961), reprinted in Flash Archives Vol. Three (DC Comics, 2002). Written by John Broome and drawn by Carmine Infantino.

CHAPTER 6

Page 70 "or of Catwoman's whip . . ." *JLA* # 13 (DC Comics, April 1998). Written by Grant Morrison and drawn by Howard Porter. The fact that the tip of Catwoman's whip is moving at roughly twice the speed of sound (the source of the loud crack it creates) was emphasized when she snapped the villain Prometheus in a particularly vulnerable area.

Page 70 *DC: The New Frontier* # 2 (DC Comics, Apr. 2004), also reprinted in DC: The New Frontier Vol. One (DC Comics, 2004). Written and drawn by Darwyn Cooke.

Page 70 *Flash* # 202 (vol. 2) (DC Comics, Nov. 2003). Written by Geoff Johns and drawn by Alberto Dose.

Page 71 "The Special Theory of Relativity can be boiled down . . ." *What Is Relativity?* L. D. Landau and G. B. Romer (Translated by N. Kemmer) (Dover, 2003).

Page 71 "with a sweeping motion . . ." *Flash* # 124 (National Comics, Nov. 1961), reprinted in Flash Archives volume 3 (DC Comics, 2002). Written by John Broome and drawn by Carmine Infantino.

Page 72 "In order for this to be true, Einstein argued . . ." *Relativity and Common Sense,* Hermann Bondi (Dover, 1980).

Page 72 "Negative Man" *My Greatest Adventure* # 80 (National Comics, June 1963), reprinted in The Doom Patrol Archives volume one (DC Comics, 2002). Written by Arnold Drake with Bob Haney and drawn by Bruno Premiani.

Page 72 "Captain Marvel" *Avengers* # 227 (Marvel Comics, Jan. 1983). Written by Roger Stern and drawn by Sal Buscema.

Page 73 *JLA* # 89 (DC Comics, late Dec. 2003). Written by Joe Kelly and drawn by Doug Mahnke.

Page 73 *Flash* # 116 (National Comics, Nov. 1960), reprinted in Flash Archives volume 2 (DC Comics, 2000). Written by John Broome and drawn by Carmine Infantino.

CHAPTER 7

Page 75 "In his first appearance . . ." *Tales to Astonish* # 27 (Marvel Comics, Jan. 1962), reprinted in Essential Ant-Man Volume One (Marvel Comics, 2002). Written by Stan Lee and Larry Lieber and drawn by Jack Kirby.

Page 75 *The Incredible Shrinking Man: A Novel,* Richard Matheson (Tor Books, 2001).

Page 76 *Tales to Astonish* # 35 (Marvel Comics, Sept. 1962), reprinted in Essential Ant-Man Volume One (Marvel Comics, 2002). Written by Stan Lee and Larry Lieber and drawn by Jack Kirby.

Page 77 "Given that ants actually communicate . . ." *Journey to the Ants,* Bert Holldobler and Edward O. Wilson (Belknap Press of Harvard University Press, 1994).

Page 77 "discussing the construction of 'time machines' " "Wormholes, Time Machines and the Weak Energy Condition," Michael S. Morris, Kip S. Thorne, and Ulvi Yurtsever, Phys. Rev. Lett. 61, 1446 (1998); "Warp Drive and Causality," Allen E. Everett, Phys. Rev. D 53, 7365 (1996); "Closed Timelike Curves Produced by Pairs of Moving Cosmic Strings: Exact

Solutions," J. Richard Gott III, Phys. Rev. Lett 66, 1126 (1991); *Black Holes and Time Warps*, K. S. Thorne (Norton, 1994).

Page 78 See the novelization *Fantastic Voyage*, Isaac Asimov (based on a screenplay by Harry Kleiner) (Bantam Books, 1966).

Page 79 "As discussed in Isaac Asimov's . . ." *Ibid*, chapter 4.

Page 80 *Fantastic Voyage II: Destination Brain*, Isaac Asimov (Doubleday, 1987).

CHAPTER 8

Page 84 "he could ride on top of an ant . . ." See, for example, *Tales to Astonish* # 35 (Marvel Comics, Sept. 1962), reprinted in Essential Ant-Man Volume 1 (Marvel Comics, 2002). Written by Stan Lee and Larry Lieber and drawn by Jack Kirby.

Page 84 "to instruct hundreds of them . . ." *Tales to Astonish* # 36 (Marvel Comics, Oct. 1962), reprinted in Essential Ant-Man Volume 1 (Marvel Comics, 2002). Written by Stan Lee and Larry Lieber and drawn by Jack Kirby.

Page 85 *Tales to Astonish* # 37 (Marvel Comics, Nov. 1962), reprinted in Essential Ant-Man Volume 1 (Marvel Comics, 2002). Written by Stan Lee and Larry Lieber and drawn by Jack Kirby.

Page 87 "an ingeneous series of levers . . ." *Intermediate Physics for Medicine and Biology* (3rd ed.), Russell K. Hobbie (American Institute of Physics, 2001); *Biomechanics of Human Motion*, M. Williams and H. R. Lissner (Saunders Press, 1962).

Page 88 "essentially the same as a fishing rod . . ." See, for example, *The Way Things Work*, David Macaulay (Houghton Mifflin, 1988) for an amusing illustration of different lever configurations.

Page 88 "The ratio of moment arms is thus 1 to 7," *Back-of-the-Envelope Physics*, Clifford Swartz (Johns Hopkins University Press, 2003).

Page 90 "What determines how high you can leap?" *On Growth and Form*, D'Arcy Thompson (Cambridge University Press, 1961).

Page 91 "It is an easy consequence of anthropomorphism," *Ibid*, page 27.

CHAPTER 9

Page 93 "Galileo was perhaps not the first person to notice . . ." *Galileo's Pendulum*, Roger G. Newton (Harvard University Press, 2004).

Page 95 "A human vocal cord is . . ." *Intermediate Physics for Medicine and Biology* (3rd ed.), Russell K. Hobbie (American Institute of Physics, 2001).

Page 97 "Alternatively, if one is too close to the source, . . ." *On Growth and Form*, D'Arcy Thompson (Cambridge University Press, 1961).

Page 97 *Atom* # 4 (National Comics, Dec./Jan. 1962), reprinted in Atom Archives Volume One (DC Comics, 2001). Written by Gardner Fox and drawn by Gil Kane.

Page 98 "An insect's eye is very good at . . ." C. J. van der Horst, "The Optics of the Insect Eye," *Acta Zool*, p. 108 (1933).

CHAPTER 10

Page 100 *Tales to Astonish* # 48 (Marvel Comics, Oct. 1963), reprinted in Essential Ant-Man Volume One (Marvel Comics, 2002). Written by Stan Lee and H. E. Huntley and drawn by Don Heck.

Page 101 *Tales to Astonish* # 49 (Marvel Comics, Nov. 1963), reprinted in Essential Ant-Man Volume One (Marvel Comics, 2002). Written by Stan Lee and drawn by Jack Kirby.

Page 101 "Yellowjacket" *Avengers* # 59 (Marvel Comics, Dec. 1968), *Avengers* # 63 (Marvel Comics, Apr. 1969), reprinted in Essential Avengers Volume Three (Marvel Comics, 2001). Written by Roy Thomas and drawn by John Buscema and Gene Colan.

Page 101 "Goliath" *Avengers* # 28 (Marvel Comics, May 1966), reprinted in Essential Avengers Vol. Two (Marvel Comics, 2000). Written by Stan Lee and drawn by Don Heck.

Page 102 *Ultimates* # 3 (Marvel Comics, May 2002). Written by Mark Millar and drawn by Bryan Hitch.

Page 102 *Fantastic Four* # 271 (Marvel Comics, Oct. 1984). Written and drawn by John Byrne.

Page 102 "Orrgo" *Strange Tales* # 90 (Marvel Comics, Nov. 1961). Written by Stan Lee and drawn by Jack Kirby.

Page 102 "Bruttu" *Tales of Suspense* # 22 (Marvel Comics, Oct. 1961). Written by Stan Lee and drawn by Jack Kirby.

Page 102 "Googam (son of Goom)" *Tales of Suspense* # 17 (Marvel Comics, May 1961). Written by Stan Lee and drawn by Jack Kirby.

Page 102 "Fin Fang Foom" *Strange Tales* # 89 (Marvel Comics, Oct. 1961). Written by Stan Lee and drawn by Jack Kirby.

Page 105 *Fantastic Four Annual* # 1 (Marvel Comics, 1963), reprinted in Marvel Masterworks: Fantastic Four Vol. Two (Marvel Comics, 2005). Written by Stan Lee and drawn by Jack Kirby.

Page 105 Footnote,
On Growth and Form, D'Arcy Thompson (Cambridge University Press, 1961).

Page 106 "The femurs of elephants" *Ultimates* # 2 (Marvel Comics, May 2002). Written by Mark Millar and drawn by Bryan Hitch.

Page 107 "If you've ever thought that the bubbles . . ." *200 Puzzling Physics Problems*, Peter Gnådig, Gyula Honyek, and Ken Riley (Cambridge University Press, 2001).

CHAPTER II

Page 111 "As you grew and matured, you needed" See, for example, *The Stuff of Life*, Eric P. Widmaier (Henry Holt and Company, 2002); *The Machinery of Life*, David S. Goodsell (Springer-Verlag, 1998); and *Stories of the Invisible*, Philip Ball (Oxford University Press, 2001).

Page 113 "When physicists studied the decay . . ." *The Elusive Neutrino: A Subatomic Detective Story*, Nickolas Solomey (W. H. Freeman & Company, 1997).

Page 114 "An automobile's efficiency . . ." *New Directions in Race Car Aerodynamics: Designing for Speed*, Joseph Katz (Bentley Publishers, 1995).

Page 116 Footnote "Positron Production in Multiphoton Light-by-Light Scattering," D. L. Burk et al., Phys. Rev. Lett. 79, 1626 (1997).

Page 118 *Flash* # 106 (DC Comics, May 1959), reprinted in Flash Archives Volume One (DC Comics, 1996). Written by John Broome and drawn by Carmine Infantino.

Page 118 *Flash* # 25 (vol. 2) (DC Comics, Apr. 1989). Written by William Messner-Loebs and drawn by Greg LaRocque.

Page 119 "physicists were confused about energy . . ." See, for example, *Warmth Disperses and Time Passes: The History of Heat*, Hans Christian von Baeyer (Modern Library, 1998).

Page 120 "At one point in *Flash* comics . . ." See, for example, *Flash* # 24 (vol. 2) (DC Comics, Apr. 1989). Written by William Messner-Loebs and drawn by Greg LaRocque.

Page 121 "consider some basic chemistry" See, for example, *The Periodic Kingdom*, P. W. Atkins (Basic Books, 1995).

Page 123 "That the mass of the resulting helium" *Just Six Numbers*, Martin Rees (Basic Books, 1999).

Page 124 "The volume of oxygen use by a runner . . ." *Energies: An Illustrated Guide to the Biosphere and Civilization*, Vaclav Smil (MIT Press, 1999).

Page 124 "the Earth's atmosphere contains . . ." There are approximately two hundred million trillion moles of gas in the Earth's atmosphere, while each mole contains Avogadro's number (0.6 trillion trillion) molecules. The Earth's atmosphere is thus estimated to contain 0.12 billion trillion trillion trillion gas molecules.

Page 125 *Flash* # 167 (DC Comics, Feb. 1967). Written by John Broome and drawn by Carmine Infantino.

Page 126 "This is one reason why golf ball have dimples." *The Physics of*

Golf, Theodore P. Jorgensen (Springer, second edition, 1999); *Golf Balls, Boomerangs and Asteroids: The Impact of Missiles on Society*, Brian H. Kaye (VCH Publishers, 1996); *500 Years of Golf Balls: History and Collector's Guide*, John F. Hotchkiss (Antique Trader Books, 1997).

Page 127 Footnote Technically, *Superman* # 130 ascribed the wrong mechanism to kryptonite's resistance to air friction when it claimed "kryptonite can't combine chemically with oxygen, which causes combustion." It may indeed not be chemically reactive with oxygen, but the heat generated when an object moves at high velocity through the atmosphere is due to the work needed to push the air molecules out of the way, and is a purely physical, rather than chemical, process.

Page 127 "The first such character" *Superman: The Complete History*, Les Daniels (Chronicle Books, 1998).

Page 127 "Not to be outdone . . ." *DC Comics: Sixty Years of the World's Favorite Comic Book Heroes*, Les Daniels (Bulfinch Press, 1995).

Page 128 "as far as most fans of the Silver Age . . ." "Comics That Didn't Really Happen," by Mark Evanier, reprinted in *Comic Books and Other Necessities of Life* (TwoMorrows Publishing, 2002).

CHAPTER 12

Page 129 *Showcase* # 34 (DC Comics, Sept./Oct. 1961), reprinted in Atom Archives Volume One (DC Comics, 2001). Written by Gardner Fox and drawn by Gil Kane.

Page 131 Fig. 17 *Atom* # 4 (DC Comics, Dec./Jan. 1962), reprinted in Atom Archives Volume One (DC Comics, 2001). Written by Gardner Fox and drawn by Gil Kane.

Page 131 *Atom* # 2 (DC Comics, Aug./Sept. 1962), reprinted in Atom Archives Volume One (DC Comics, 2001). Written by Gardner Fox and drawn by Gil Kane.

Page 132 "The field of thermodynamics . . ." *Warmth Disperses and Time Passes: The History of Heat*, Hans Christian von Baeyer (Modern Library, 1998); *A Matter of Degrees*, Gino Segre (Penguin Books, 2002).

Page 135 "Another example:" *Energies: An Illustrated Guide to the Biosphere and Civilization*, Vaclav Smil (MIT Press, 1999).

Page 136 "this concept called 'entropy,' is . . ." *Understanding Thermodynamics*, H. C. Van Ness (Dover Publications, 1969).

Page 139 *West Coast Avengers* # 42 (Marvel Comics, Mar. 1989). Written and drawn by John Byrne.

Page 140 "Could I use the talents of the Atom . . ." *Warmth Disperses and Time Passes: The History of Heat*, Hans Christian von Baeyer (Modern Library, 1998).

Page 143 "radio-wave background radiation" *Temperatures Very Low and Very High*, Mark W. Zemansky (Dover Books, 1964).

Page 144 "Many of the elder statesmen of physics" *Philosophy of Science: The Historical Background,* Joseph J. Kockelmans (ed.) (Transaction Publishers, 1999).

Page 144 Planck quote *Scientific Autobiography and Other Papers,* Max K. Planck (translated by F. Gaynor) (Greenwood Publishing Group, 1968).

Page 144 "A key development" *An Introduction to Stochastic Processes in Physics,* Don S. Lemons (Johns Hopkins University Press, 2002).

Page 144 "it was not until 1905 . . ." *Investigations of the Theory of the Brownian Movement,* Albert Einstein (Dover, 1956).

Page 145 "The random collisions of the air on our eardrums . . ." "How the Ear's Works Work," A. J. Hudspeth, *Nature* 341, 397 (1989); "Brownian Motion and the Ability to Detect Weak Auditory Signals," I. C. Gebeshuber, A. Mladenka, F. Rattay, and W. A. Svrcek-Seiler, *Chaos and Noise in Biology and Medicine,* ed. C. Taddei-Ferretti (World Scientific, 1998). Note that this is not the high-pitch tone that many of us hear. That high-frequency sound is most likely tinnitus, resulting from damage (either from loud noises or old age) to the cilia that detect sound waves in the inner ear.

CHAPTER 13

Page 146 "Stan Lee, head writer and editor, . . ." *Excelsior!: The Amazing Life of Stan Lee,* Stan Lee and George Mair (Fireside, 2002).

Page 146 *X-Men* # 1 (Marvel Comics, Sept. 1963), reprinted in Marvel Masterworks: X-Men Vol. One (Marvel Comics, 2002). Written by Stan Lee and drawn by Jack Kirby.

Page 147 *X-Men* # 47 (Marvel Comics, Aug. 1968). Written by Arnold Drake and drawn by Werner Roth.

Page 147 *X-Men* # 8 (Marvel Comics, Nov. 1964), reprinted in Marvel Masterworks: X-Men Vol. One (Marvel Comics, 2002). Written by Stan Lee and drawn by Jack Kirby.

Page 147 "A snowflake is created when . . ." *The Snowflake: Winter's Secret Beauty,* Kenneth G. Libbrecht and Patricia Rasmussen (Voyageur Press, 2003).

Page 148 "Einstein's equation for how far a fluctuating atom . . ." *Investigations of the Theory of the Brownian Movement,* Albert Einstein (Dover, 1956).

Page 148 "The exact details . . ." "Instabilities and Pattern Formation in Crystal Growth," J. S. Langer, *Reviews of Modern Physics* 52, 1 (1980).

Page 150 Fig. 19 *Amazing Spider-Man* # 92 (Marvel Comics, Jan. 1971), reprinted in Spider-Man: The Death of Captain Stacy (Marvel Comics, 2004). Written by Stan Lee and drawn by Gil Kane and John Romita.

Page 151 *Giant-Sized X-Men* # 1 (Marvel Comics, 1975), reprinted in

Marvel Masterworks: Uncanny X-Men Vol. One (Marvel Comics, 2003). Written by Len Wein and drawn by David Cockrum.

Page 152 "At its core, the weather . . ." *The Essence of Chaos*, Edward Lorenz (University of Washington Press, 1996); *The Coming Storm*, Mark Masline (Barron's, 2002).

Page 154 Fig. 21 X-Men # 145 (Marvel Comics, May 1981). Written by Chris Claremont and drawn by Dave Cockrum and Joe Rubinstein.

Page 155 "A final thought . . ." *Lord Kelvin and the Age of the Earth*, Joe D. Burchfield (University of Chicago Press, 1990); *Degrees Kelvin*, David Lindley (Joseph Henry Press, 2004).

CHAPTER 14

Page 158 *Tales of Suspense* # 39 (Marvel Comics, Mar. 1963), reprinted in Marvel Masterworks: The Invincible Iron Man Vol. 1 (Marvel Comics, 2003). Written by Stan Lee and Larry Lieber and drawn by Don Heck.

Page 159 "When the Melter first appeared" *Tales of Suspense* # 47 (Marvel Comics, Nov. 1963), reprinted in Marvel Masterworks: The Invincible Iron Man Vol. 1 (Marvel Comics, 2003). Written by Stan Lee and drawn by Steve Ditko.

Page 159 "When this happens, a chemical bond forms . . ." *The Periodic Kingdom*, P. W. Atkins (Basic Books, 1995).

Page 160 "What determines the exact temperature and pressure . . ." *Gases, Liquids and Solids*, D. Tabor (Cambridge University Press, 1979).

Page 160 "In a conventional oven . . ." *On Food and Cooking*, Harold McGee (Scribner, revised and updated edition, 2004); *The Science of Cooking*, Peter Barham (Spring, 2001).

Page 164 *Tales of Suspense* # 90 (Marvel Comics, Jun. 1967), Essential Iron Man Vol. Two (Marvel Comics, 2004). Written by Stan Lee and drawn by Gene Colan.

Page 164 "Such a microwave-based 'heat ray' that . . ." "Report: Raytheon 'heat beam' weapon ready for Iraq," *Boston Business Journal*, Dec. 1, 2004.

CHAPTER 15

Page 166 Footnote *West Coast Avengers* #13, vol. 2 (Marvel Comics, Oct. 1986). Written by Steve Englehart and drawn by Al Milgrom.

Page 167 *Adventure* # 247 (National Comics, April 1958), reprinted in Legion of Superheroes Archives Vol. One (DC Comics, resissue edition, 1991). Written by Otto Binder and drawn by Al Plastino.

Page 168 *Amazing Spider-Man* # 9 (Marvel Comics, Feb. 1963), reprinted in Marvel Masterworks: Amazing Spider-Man Vol. One (Marvel Comics, 2003). Written by Stan Lee and drawn by Steve Ditko.

Page 170 "is approximately the same size as a carbon atom . . ." *Back-of-the-Envelope Physics*, Clifford Swartz (Johns Hopkins University Press, 2003).

Page 172 "evil Wizard's anti-gravity discs" First seen in *Strange Tales* # 118 (Marvel Comics, March 1964), reprinted in Essential Human Torch Vol. One (Marvel Comics, 2003). Written by Stan Lee and drawn by Dick Ayers.

Page 173 *Flash* # 208, vol. 2 (DC Comics, May 2004). Written by Geoff Johns and drawn by Howard Porter.

Page 173 "George de Mestral's investigations . . ." *Why Didn't I Think of That?*, Allyn Freeman and Bob Golden (John Wiley and Sons, 1997).

Page 174 "Evidence for Van der Waals Adhesion in Gecko Setae," K. Autumn, M. Sitti, Y. A. Liang, A. M. Peattie, W. R. Hansen, S. Sponberg, T. W. Kenny, R. Fearing, J. N. Israelachvili, and R. J. Full, *Proc. National Acad. Sciences* 99, 12,252 (2002).

Page 174 "development of 'gecko tape,'" "Microfabricated Adhesive Mimicking Gecko Foot-Hair," A. K. Geim, S. V. Dubonos, 2, I. V. Grigorieva, K. S. Novoselov, A. A. Zhukov, and S. Yu. Shapoval, Nature Materials 2, 461 (2003).

CHAPTER 16

Page 178 *Superman* # 1 (National Comics, June 1939), reprinted in Superman Archives Volume 1 (DC Comics, 1989). Written by Jerry Siegel and drawn by Joe Shuster.

Page 180 *Amazing Spider-Man* # 9 (Marvel Comics, Feb. 1963), reprinted in Marvel Masterworks: Amazing Spider-Man Vol. One (Marvel Comics, 2003). Written by Stan Lee and drawn by Steve Ditko.

Page 180 *Amazing Spider-Man* Annual # 1 (Marvel Comics, Feb. 1964), reprinted in Marvel Masterworks: Amazing Spider-Man Vol. Two (Marvel Comics, 2002). Written by Stan Lee and drawn by Steve Ditko.

Page 181 "a comic book writer would generate a script . . ." *Man of Two Worlds, My Life in Science Fiction and Comics*, Julius Schwartz with Brian M. Thomsen (HarperEntertainment, New York), 2000.

Page 181 "in 1965, to pick a particular year . . ." *Comic Book Marketplace* # 99 (Gemstone Publishing, Feb. 2003).

Page 182 "With so many stories being created every month . . ." *Stan Lee and the Rise and Fall of the American Comic Book*, Jordan Raphael and Tom Spurgeon (Chicago Review Press, 2003); *Tales to Astonish: Jack Kirby, Stan Lee and the American Comic Book Revolution* by Ronin Ro (Bloomsbury, 2004).

Page 183 *Daredevil* # 2 (Marvel Comics, June 1964), reprinted in Marvel

Masterworks: Daredevil Vol. One (Marvel Comics, 2004). Written by Stan Lee and drawn by Joe Orlando.

CHAPTER 17

Page 184 "a perfect illustration of a mysterious . . ." *Amazing Spider-Man* # 9 (Marvel Comics, Feb. 1963), reprinted in Marvel Masterworks: Amazing Spider-Man Vol. One (Marvel Comics, 2003). Written by Stan Lee and drawn by Steve Ditko.

Page 185 "This phenomenon, termed the Ampere effect, . . ." *Electric Universe: The Shocking True Story of Electricity,* David Bodanis (Crown, 2005).

Page 185 *Daredevil* # 2 (Marvel Comics, June 1964), reprinted in Marvel Masterworks: Daredevil Vol. One (Marvel Comics, 2004). Written by Stan Lee and drawn by Joe Orlando.

Page 186 "I'll use a nice argument . . ." *Discovering the Natural Laws: The Experimental Basis of Physics,* Milton A. Rothman (Dover Press, 1989).

Page 187 "The test charge therefore sees . . ." *Electricity and Magnetism—Berkeley Physics Course Vol. 2,* Edward M. Purcell (McGraw Hill, 1963).

Page 188 *Superboy* # 1 (National Comics, Mar.–Apr. 1949). Written by Edmond Hamilton and drawn by John Sikela and Ed Dobrotka.

CHAPTER 18

Page 191 *X-Men* # 1 (Marvel Comics, Sept. 1963), reprinted in Marvel Masterworks: X-Men Vol. One (Marvel Comics, 2002). Written by Stan Lee and drawn by Jack Kirby.

Page 192 Footnote *Atom* # 3 (DC Comics, Oct.–Nov. 1962), reprinted in Atom Archives Vol. One (DC Comics, 2001). Written by Gardner Fox and drawn by Gil Kane.

Page 193 "Hemoglobin is a very large molecule . . ." *The Machinery of Life,* David S. Goodsell (Springer-Verlag, 1998).

Page 195 Footnote I thank Prof. E. Dan Dahlberg of the University of Minnesota and Dr. Roger Proksh of Asylum Research for demonstrating this low-tech "magnetic force microscope."

Page 195 "Materials that form magnetic domains . . ." *Magnets: The Education of a Physicist,* Francis Bitter (Doubleday, 1959).

Page 196 "It is through our diamagnetism" "Everyone's Magnetism," Andrey Geim, *Physics Today* 51, p. 36 (Sept. 1998); "Magnet levitation at your fingertips," A. K. Geim, M. D. Simon, M. I. Boamfa, and L. O. Heflinger. *Nature* 400, p. 323 (1999).

Page 196 See the web page for the High Field Magnetic Laboratory at the University of Nijmegen in the Netherlands: http://www.hfml.ru.nl/levitate.html for some great images of levitating objects.

Page 198 "Magnetism is, at its heart, . . ." *Discovering the Natural Laws: The Experimental Basis of Physics,* Milton A. Rothman (Dover Press, 1989).

Page 200 *The Dark Knight Strikes Again* # 1 (DC Comics, 2001). Reprinted in The Dark Knight Strikes Again (DC Comics, 2003). Written and drawn by Frank Miller.

Page 200 "All commercial power plants . . ." *Energy: Its Use and the Environment,* Roger A. Hinrichs and Merlin Kleinbach (Brooks/Cole, 2002), Third Edition.

CHAPTER 19

Page 202 "help keep comic-book publishers solvent" *Seal of Approval, The History of the Comics Code,* Amy Kiste Nyberg (University of Mississippi Press, Jackson, Mississippi), 1998.

Page 202 Western comics at DC and Marvel . . . *Comic Book Culture: An Illustrated History,* Ron Goulart (Collectors Press Inc., 2000); *DC Comics: Sixty Years of the World's Favorite Comic Book Heroes, Les Daniels* (Bulfinch Press, 1995); *Marvel: Five Fabulous Decades of the World's Greatest Comics, Les Daniels* (Harry Abrams, 1991).

Page 202 "It was the Scottish physicist . . ." *The Man Who Changed Everything: The Life of James Clerk Maxwell,* Basil Mahon (John Wiley & Sons, 2003).

Page 205 "600 million tons of hydrogen nuclei every second . . ." "The Evolution and Explosion of Massive Stars," S. E. Woolsey and A. Heger, *Rev. Modern Physics* 74, p. 1015 (Oct. 2002).

Page 206 "light generated from a nuclear fusion reaction . . ." "How Long Does It Take for Heat to Flow Through the Sun?" G. Fiorentini and B. Rici, *Comments on Modern Physics* 1, p. 49 (1999).

Page 207 "While his shattered spine may have left him . . ." *X-Men* # 20 (Marvel Comics, May 1966), reprinted in Marvel Masterworks: X-Men Vol. Two (Marvel Comics, 2003). Written by Roy Thomas and Drawn by Jay Gavin.

Page 207 "The role of nerve cells . . ." *Synaptic Self: How Our Brains Become Who We Are,* Joseph LeDoux (Penguin, 2002); *I of the Vortex,* R. R. Llinas (MIT Press, 2001).

Page 209 *X-Men* # 7 (Marvel Comics, Sept. 1964), reprinted in Marvel Masterworks: X-Men Vol. 1 (Marvel Comics, 2002). Written by Stan Lee and Drawn by Jack Kirby.

Page 209 "Television signals consist of . . ." See *The Way Things Work,*

David Macaulay (Houghton Mifflin Company, 1988), for an accessible, graphical illustration of the mechanisms underlying television broadcasts and reception, and *How Does It Work?* Richard M. Koff (Signet, 1961), for a more technical discussion.

Page 210 "A sensitive antenna placed near this monitor . . ." "Electromagnetic Radiation from Video Display Units: An Eavesdropping Risk?," Wim Van Eck, *Computers and Security* 4, p. 269 (1985).

Page 211 "Neuroscientists have developed a research tool . . ." "Experimentation with a Transcranial Stimulation System for Functional Brain Mapping," G. J. Ettinger, W. E. L. Grimson, M. E. Leventon, R. Kikinis, V. Gugino, W. Cote et al. *Med. Image Analysis* 2, p. 133 (1998); "Transcranial Magnetic Stimulation and the Human Brain," M. Hallett, *Nature* 406, p. 147 (2000). A technical overview can be found in *Transcranial Magnetic Stimulation: A Neurochronometrics of Mind*, Vincent Walsh and Alvaro Pascual-Leone (MIT Press, 2003).

CHAPTER 20

Page 215 *Fantastic Four* # 5 (Marvel Comics, July 1962), reprinted in Marvel Masterworks: Fantastic Four Vol. One (Marvel Comics, 2003). Written by Stan Lee and drawn by Jack Kirby.

Page 216 *Fantastic Four* # 10 (Marvel Comics, Jan. 1963), reprinted in Marvel Masterworks: Fantastic Four Vol. One (Marvel Comics, 2003). Written by Stan Lee and drawn by Jack Kirby.

Page 216 *Fantastic Four* # 16 (Marvel Comics, July 1963), reprinted in Marvel Masterworks: Fantastic Four Vol. Two (Marvel Comics, 2005). Written by Stan Lee and drawn by Jack Kirby.

Page 216 *Fantastic Four* # 76 (Marvel Comics, July 1968), reprinted in Marvel Masterworks: Fantastic Four Vol. Eight (Marvel Comics, 2005). Written by Stan Lee and drawn by Jack Kirby.

Page 217 *Atom* # 5 (DC Comics, Feb./Mar. 1963), reprinted in Atom Archives Vol. One (DC Comics, 2001). Written by Gardner Fox and drawn by Gil Kane; *Atom* # 4 (DC Comics, Dec./Jan. 1962), reprinted in Atom Archives Vol. One (DC Comics, 2001). Written by Gardner Fox and drawn by Gil Kane; *Atom* # 19 (DC Comics, Jun./Jul. 1965), reprinted in JLA: Zatanna's Search (DC Comics, 2004). Written by Gardner Fox and drawn by Gil Kane; *Justice League of America* # 18 (DC Comics, Mar. 1963), reprinted in Justice League of America Archives Vol. Three (DC Comics, 1994). Written by Gardner Fox and drawn by Mike Sekowsky; *Brave and the Bold* # 53 (DC Comics, Apr.–May, 1964). Written by Bob Haney and drawn by Alexander Toth.

Page 218 "At the end of the nineteenth century, . . ." *Thirty Years That Shook Physics: The Story of Quantum Theory*, G. Gamow (Dover, 1985).

Page 219 "This is how the surface temperature of the sun . . ." *Temperatures Very Low and Very High*, Mark W. Zemansky (Dover, 1964).

Page 221 "The fact that the energy of electrons . . ." *The New World of Mr. Tompkins*, G. Gamow and R. Stannard (Cambridge University Press, 1999).

Page 224 "Imagine an electron orbiting a nucleus . . ." *The Quantum World*, J. C. Polkinghorne (Princeton University Press, 1984).

Page 226 "The lighter-than-air element Helium . . ." *Helium: Child of the Sun*, Clifford W. Seibel (University Press of Kansas, 1968).

CHAPTER 21

Page 229 *Showcase* # 4 (National Comics, Oct. 1956), reprinted in Flash Archives Volume One (DC Comics, 1996). Written by Robert Kanigher and drawn by Carmine Infantino.

Page 229 *Flash* # 123 (DC Comics, Sept. 1961), reprinted in Flash Archives Vol. Three (DC Comics, 2002). Written by Gardner Fox and drawn by Carmine Infantino.

Page 231 "The crossover meeting between the Silver Age and Golden Age Flash . . ." *Man of Two Worlds, My Life in Science Fiction and Comics*, Julius Schwartz with Brian M. Thomsen (HarperEntertainment, 2000).

Page 231 "So popular was this meeting of the two Super-teams . . ." See, for example, *Crisis on Multiple Earths* Volumes One, Two, Three (DC Comics, 2002, 2003, 2004).

Page 231 "Billy Batson, who could become a superhero by shouting 'Shazam!'" *DC Comics: Sixty Years of the World's Favorite Comic Book Heroes*, Les Daniels (Bulfinch Press, 1995).

Page 232 "The yearlong miniseries" *Crisis on Infinite Earths* (DC Comics, 2000). Written by Marv Wolfman and drawn by George Perez.

Page 233 "more like those described in the Marvel comic universe" *What If Classics* (Marvel Comics, 2004).

Page 233 "After a great deal of effort" *Thirty Years That Shook Physics: The Story of Quantum Theory*, G. Gamow (Dover Press, 1985).

Page 234 "two points about mathematics" *Euclid's Window: The Story of Geometry from Parallel Lines to Hyperspace*, Leonard Mlodinow (Touchstone, 2001).

Page 235 *My Greatest Adventures* # 80 (DC Comics, June 1963), reprinted in Doom Patrol Archives Vol. One (DC Comics, 2002). Written by Arnold Drake and drawn by Bruno Premiani.

Page 235 *X-Men* # 1 (Marvel Comics, Sept. 1963), reprinted in Marvel Masterworks: X-Men Vol. One (Marvel Comics, 2002). Written by Stan Lee and drawn by Jack Kirby.

Page 236 "research of comic-book historians" See, for example, *Comic Book Marketplace* # 64 (Gemstone Publications, Nov. 1998).

Page 236 "Another publishing synchronicity" *Back Issue* # 6 (TwoMorrows Publishing, Oct. 2004).

Page 237 "While we may not know how Schrödinger" *Schrödinger: Life and Thought,* Walter Moore (Cambridge University Press, 1989).

Page 238 "Given that the *average* values are the only quantities . . ." The notion that quantum mechanics is a complete theory that always provides accurate predictions of experimental observations, but does not necessarily describe an external reality, is not universally accepted among physicists. The growing body of experiments on macroscopic quantum behavior would, however, tend to support this interpretation (see "The Quantum Measurement Problem," A. J. Leggett, *Science* 307, p. 871 (2005)).

Page 240 "They posed the following situation . . ." *Schrödinger's Rabbits: The Many Worlds of Quantum,* Colin Bruce (Joseph Henry Press, 2004).

Page 241 *JLA* # 19 (DC Comics, June 1998). Written by Mark Waid and drawn by Howard Porter.

Page 241 "In 1957 Hugh Everett III argued . . ." *The Many-Worlds Interpretation of Quantum Mechanics,* edited by Bryce S. DeWitt and Neill Graham (Princeton University Press, 1973). Contains a reprint of Everett's Ph.D. thesis and a longer discussion of his ideas, along with articles by other physicists. Interestingly, Everett refered to his theory as involving "Relative States," and it was DeWitt who coined the expression "Many Worlds."

Page 241 Footnote *Animal Man* # 32 (DC Comics, Feb. 1991). Written by Peter Milligan and drawn by Chas Troug.

Page 243 "A gross oversimplification of string theory . . ." An excellent introduction to String Theory can be found in *The Elegant Universe,* Brian Greene (W. W. Norton, 1999).

Page 243 Footnote *Strange Tales* # 129 (Marvel Comics, Feb. 1965), reprinted in Essential Dr. Strange Vol. One (Marvel Comics, 2001). Written by Don Rico and drawn by Steve Ditko.

Page 243 "This may be dangerous; . . ." See *The Pleasure of Finding Things Out,* Richard P. Feynman (Perseus Books, 1999).

Page 244 "Physicists developing quantum gravity . . ." *The Fabric of Reality: The Science of Parallel Universes and Its Implications,* David Deutsch (Penguin, 1997); *Parallel Worlds,* Michio Kaku (Doubleday, 2005).

Page 244 "Recently some scientists have claimed that time travel . . ." *The Future of Spacetime,* Stephen W. Hawking, Kip S. Thorne, Igor Novikov, Timothy Ferris, and Alan Lightman (W. W. Norton and Company, 2002); *Time Travel in Einstein's Universe,* J. Richard Gott (Mariner Books, 2001).

Page 244 *Superman* # 146 (DC Comics, July 1961). Written by Jerry Siegel and drawn by Al Plastino.

Page 248 *Avengers* # 267 (Marvel Comics, May 1986), reprinted in Avengers: Kang—Time and Time Again (Marvel Comics, 2005). Written by Roger Stern and drawn by John Buscema and Tom Palmer.

CHAPTER 22

Page 250 Fig. 32 *X-Men* # 130 (Marvel Comics, Feb. 1980). Written by Chris Claremont and drawn by John Byrne.

Page 250 Fig. 33 *Flash* # 123 (DC Comics, Sept. 1961), reprinted in Flash Archives Vol. Three (DC Comics, 2002). Written by Gardner Fox and drawn by Carmine Infantino.

Page 250 "This is an intrinsically quantum mechanical phenomenon . . ." A mathematical discussion of this phenomenon can be found in *Quantum Theory of Tunneling*, Mohsen Razavy (World Scientific, 2003).

Page 252 "Scanning Tunneling Microscope" "The Scanning Tunneling Microscope," G. Binnig and H. Rohrer, *Scientific American* 253, p. 40 (1985); "Vacuum tunneling: A new technique for microscopy." C. F. Quate, *Physics Today* 39, p. 26 (1986); *Solid State Electronic Devices* (5th ed.), Ben G. Streetman and Sanjay Banerjee (Prentice Hall, 2000).

Page 255 *X-Men* # 141 (Marvel Comics, Jan. 1981), reprinted in Days of Future Past (Marvel Comics, 2004). Written by Chris Claremont and drawn by John Byrne.

Page 255 *Astonishing X-Men* # 4 (Marvel Comics, Oct. 2004). Written by Joss Whedon and drawn by John Cassaday.

CHAPTER 23

Page 257 "The solid-state transistor is the fountainhead . . ." *Crystal Fire: Birth of the Information Age*, Michael Riordan (Norton, 1997).

Page 258 *Showcase* # 22 (DC Comics, Oct. 1959), reprinted in Green Lantern Archives Vol. One (DC Comics, 1998). Written by John Broome and drawn by Gil Kane.

Page 258 *Showcase* # 6 (National Comics, Jan.–Feb. 1957), reprinted in Challengers of the Unknown Archives Vol. One (DC Comics, 2003). Written by Dave Wood and drawn by Jack Kirby.

Page 258 "The Marvel Age of Comics began . . ." *Marvel Comics Presents Fantastic Firsts* (Marvel Comics, 2001).

Page 258 *Tales of Suspense* # 39 (Marvel Comics, Mar. 1963), reprinted in Marvel Masterworks: The Invincible Iron Man Vol. 1 (Marvel Comics, 2003). Written by Stan Lee and Larry Lieber and drawn by Don Heck.

Page 259 *Iron Man* # 144 (Marvel Comics, Mar. 1981). Written by David Michelinie and drawn by Joe Brozowski and Bob Layton.

Page 259 "And boy did Shellhead . . ." See Essential Iron Man Vol. One and Two (Marvel Comics, 2000, 2004).

Page 260 "The weapons that were distributed . . ." *Tales of Suspense* # 55 (Marvel Comics, Jul. 1963), reprinted in Essential Iron Man Vol. One (Marvel Comics, 2000). Written by Stan Lee and Larry Lieber and drawn by Don Heck.

Page 264 *Iron Man* # 132 (Marvel Comics, Mar. 1980). Written by David Michelinie and drawn by Jerry Bingham and Bob Layton.

Page 266 "the phenomenon of diamagnetic levitation . . ." See the web page for the High Field Magnetic Laboratory at the University of Nijmegen in the Netherlands: http://www.hfml.ru.nl/levitate.html for some great images of levitating objects.

Page 267 "hand-held pulsed-energy weapons . . ." "Star Wars Hits the Streets," David Hambling, *New Scientist,* issue no. 2364 (October 12, 2002).

Page 271 "Semiconductor devices are typically constructed . . ." *Quantum Electronics,* John R. Pierce (Doubleday Anchor, 1966).

Page 273 "we all possess invisible cells . . ." "Dying to See," Ralf Dahm, *Scientific American* 291, p. 83 (Oct. 2004); "Lens Organelle Degradation," Steven Bassnett, *Experimental Eye Research* 74, p. 1 (2002).

Page 273 *Fantastic Four* # 62, vol. 3 (Marvel Comics, Dec. 2002). Written by Mark Waid and drawn by Mike Wieringo.

Page 277 "on the day Bardeen learned . . ." *The Chip: How Two Americans Invented the Microchip and Launched a Revolution,* T. R. Reid (Simon & Shuster, 1985).

CHAPTER 24

Page 281 "The first young mutant . . ." See, "Call Him . . . Cyclops!" in X-Men # 43 (Marvel Comics, Apr. 1968). Written by Roy Thomas and drawn by Werner Roth.

Page 283 "in the early days of the Golden Age . . ." *Superman* # 1 (National Comics, June 1939), reprinted in Superman Archives Volume 1 (DC Comics, 1989). Written by Jerry Siegel and drawn by Joe Shuster.

Page 284 "Before long he was lifting . . ." See Superman: The Man of Tomorrow Archives Vol. One (DC Comics, 2004) for a selection of feats of super-strength.

Page 284 "even hold up a mountain . . ." *Secret Wars* # 4 (Marvel Comics, Aug. 1984), reprinted in Marvel Super Heroes Secret Wars (Marvel Comics, 2005). Written by Jim Shooter and drawn by Bob Layton.

Page 284 *World's Finest* # 86, (National Comics, Jan.–Feb. 1957), reprinted in World's Finest Comics Archives Vol. Two (DC Comics, 2001). Written by Edmond Hamilton and drawn by Dick Sprang.

Page 286 *Fantastic Four* # 249 (Marvel Comics, Dec. 1982), reprinted in Essential John Byrne Vol. Two (Marvel Comics, 2004). Written and drawn by John Byrne.

Page 286 "two mathematicians, Euler and LaGrange, proved . . ." Euler, *Acta Acad. Sci. Imp. Petropol.,* pp. 163–193 (1778); G. Greenhill, *Proc. Camb. Phil. Soc.* 4, p. 65 (1881); *On Growth and Form,* D'Arcy Thompson (Cambridge University Press, 1961).

Page 287 "Just such a fate inevitably befell Stilt-Man . . ." *Daredevil* # 8 (Marvel Comics, June 1965), reprinted in Essential Daredevil Vol. One (Marvel Comics, 2002). Written by Stan Lee and drawn by Wallace Wood. *Daredevil* # 26 (Marvel Comics, Mar. 1967), reprinted in Essential Daredevil Vol. Two (Marvel Comics, 2004). Written by Stan Lee and drawn by Gene Colan. *Daredevil* # 48 (Marvel Comics, Jan. 1969), reprinted in Essential Daredevil Vol. Two (Marvel Comics, 2004). Written by Stan Lee and drawn by Gene Colan.

Page 287 "Doctor Octopus, is able to walk." See *Amazing Spider-Man* # 3, 11, 12 (Marvel Comics, July 1963, Apr. 1964, May 1964), reprinted in Essential Spider-Man Vol. One (Marvel Comics, 2002). Written by Stan Lee and drawn by Steve Ditko.

Page 287 *JLA* # 58 (DC Comics, Nov. 2001). Written by Mark Waid and drawn by Mike Miller.

Page 290 "A common misconception is that the pressure change . . ." K. Weltner, *American Journal of Physics* 55, pp. 50–54 (1987).

Page 291 "Birds such as the California condor . . ." "The Simple Science of Flight: From Insects to Jumbo Jets," Henk Tennekes (MIT Press, 1997).

Page 292 *Avengers* # 57 (Marvel Comics, Oct. 1967), reprinted in Essential Avengers Vol. Three (Marvel Comics, 2001). Written by Roy Thomas and drawn by John Buscema.

Page 294 *Power of the Atom* # 12 (DC Comics, May 1989). Written by William Messner-Loebs and drawn by Graham Nolan.

Page 294 *Showcase* # 34 (DC Comics, Sept./Oct. 1961), reprinted in Atom Archives Volume One (DC Comics, 2001). Written by Gardner Fox and drawn by Gil Kane.

Page 295 "When you speak, complex sound waves . . ." *The Way Things Work*, David Macaulay (Houghton Mifflin Company, 1988) has an accessible description of the physics underlying telephones. See also *How Does It Work?* by Richard Mikoff (Signet, 1961).

Page 297 "When a low mass star . . ." *Astronomy. The Solar System and Beyond* (2nd edition), Michael A. Seeds (Brooks/Cole, 2001).

Page 298 "If, as some astrophysicists have suggested . . ." "Disks of Destruction," Robert Irion, *Science* 307, pp. 66–67 (2005).

AFTERWORD

Page 301 " *Fantastic Four* # 22 (Marvel Comics, Jan. 1964). Reprinted in Marvel Masterworks: Fantastic Four Vol. Three (Marvel Comics, 2003). Written by Stan Lee and drawn by Jack Kirby.

Page 301 Footnote *Green Lantern* # 24 (DC Comics, Oct. 1963), reprinted in Green Lantern Archives Vol. Four (DC Comics, 2002). Written by John Broome and drawn by Gil Kane.

Page 301 *JLA* # 19 (DC Comics, June 1998). Written by Mark Waid and drawn by Howard Porter.

Page 302 "turned him into a human marionette." *Flash* # 133 (DC Comics, Dec. 1962). Written by John Broome and drawn by Carmine Infantino.

Page 302 "if you could travel one thousand years into the past . . ." In a *What If* tale *What If* # 33 (Marvel Comics, Jun. 1982), written by Steven Grant and drawn by Don Perlin, Iron Man is trapped back in the days of King Arthur by a double-crossing Doctor Doom. With no way to return to the present, he employs his twentieth-century knowledge of science and engineering to usher in a millennium of world-wide peace and prosperity.

Page 303 "Knowledge is itself the basis of civilization." "To the United Nations," Niels Bohr, *Impact of Science on Society* 1, p. 68 (1950).

Page 303 "A New Model Army Soldier Rolls Closer to the Battlefield," Tim Weiner, *New York Times*, Feb. 16, 2005; "Who Do You Trust: G.I. Joe or A.I. Joe?," George Johnson, *New York Times*, Ideas and Trends, Feb. 20, 2005.

Page 304 *Superman* # 156 (DC Comics, Oct. 1962), reprinted in Superman in the Sixties (DC Comics, 1999). Written by Edmond Hamilton and drawn by Curt Swan.

Page 304 "Face Front, True Believer," à la Stan Lee, in practically every Marvel Comic in the 1960s.

ACKNOWLEDGMENTS

THE SEED OF THE IDEA for the freshman semi-nar on the physics of superheroes that preceded the writing of this book was unintentionally planted by Prof. Terry Jones of the as-tronomy department at the University of Minnesota during a stu-dent's preliminary oral exam. A traditional academic hurdle in many physics graduate programs involves an examination of the students' general knowledge of physics, administered by faculty members who direct their questions to a student, who is armed only with chalk and a blackboard, and must answer in real time. Terry's question: "How much energy would it take for the Death Star to blow up the planet Alderaan in *Star Wars (Episode IV—A New Hope)*?" led me to think of other exploding planets, and in-spired me to develop my freshman seminar on superhero science.

Speaking of inspiration, I owe a debt to my many physics teach-ers and mentors at the college, graduate, and post-graduate levels. In particular I would like to thank Steve Cotsalas, John Jacobson, Peter Tea, Robert Alfano, Narkis Tzoar, Timothy Boyer, Frederick W. Smith, Kenneth Rubin, Sidney R. Nagel, Robert A. Street, and Hellmut Fritzsche. They taught me physics and, by their example, how to be a physicst.

I am also grateful to the creators of the many comic book adven-tures I have enjoyed over these many years. There are too many names to list them all, but I am especially appreciative of the efforts of the comic book creators of my youth: Gardner Fox, John Broome, Carmine Infantino, Gil Kane, Gene Colan, John Romita, Robert Kanigher, Steve Ditko, and the big three, Julius Schwartz, Stan Lee, and Jack Kirby. Their stories, featuring heroes that employed their intelligence as well as their superpowers to save the day, taught me an early lesson on the importance of "brain power," though a power ring can be handy also.

I would like to thank Prof. Lawrence M. Krauss for graciously agreeing to write the foreword for this book. I would also like to take this opportunity to acknowledge Craig Shutt (Mr. Silver Age), from whose book *Baby Boomer Comics* I shamelessly lifted the joke in the afterword's title (quoting a Stan Lee *Fantastic Four* story, who in turn got it from the Bible).

I am grateful to my mom for instilling in me a love of reading and setting an example of lifelong learning and critical thinking. My own children, Thomas, Laura, and David, have served as willing test subjects for many of the arguments presented here, and I benefited from their feedback. I also thank Laura Adams and Allen Goldman for the STM image in figure 35. My thanks to friends, family, and employees at Dreamhaven Books and Comics for support and advice.

I could not have written this book without the input from the students in my freshman seminar course from 2001 through 2003, and the evening Compleat Scholar class I taught in 2003. They enriched the class with their insightful comments, clever ideas, and unique perspectives on superhero physics. In particular, questions raised by Eric Caron, Kristin Barbieri, Matt Bialick, Drew Goebel, and Christopher Brummund directly inspired the discussion of some of the topics covered in this book.

One of the bonuses of writing a book on the physics of superheroes is that it has provided a new perspective from which old problems can be viewed. However, a drawback to seriously considering fantastic topics such as running at super speed or being able to adjust one's tunneling probability at will is that one cannot rely on experiment as a check on analysis. This handicap was mitigated by the thoughtful consideration provided by my colleagues in the School of Physics and Astronomy at the University of Minnesota. I am extremely grateful to Prof. E. Dan Dahlberg, who generously volunteered his time to carefully read the entire manuscript in draft form, and caught many mistakes or oversimplifications. In addition, Professors Benjamin Bayman, Charles E. Campbell, Michel Janssen, Russell Hobbie, Marco Peloso, and John Broadhurst reviewed selected chapters of this book—the book is much improved for their input. I also benefited from the helpful comments and suggestions of Mark Waid, Gerard Jones, and Kurt Busiek. Any errors or confusion that remain are solely my responsibility, though I reserve the right to claim that they have been deliberately left in the text as "Easter eggs" for the attentive reader to discover.

Finally, I would like to offer thanks and acknowledgment for the contributions of the following individuals—without their assistance this book would have been much poorer: my agent, Jay Mandel, for wondering whether there might be a book in this superhero approach to teaching physics, for crucial guidance in the early drafts, and for helping to set the tone of the book. I am very fortunate to have Brendan Cahill of Gotham Books for an editor. His technical advice for structuring the book greatly improved the manuscript and saved this first-time author from several "freshman mistakes." Brendan's vision for this book complemented my own, and—just as importantly—his own knowledge of superhero comic books enabled him to suggest examples that I hadn't considered. In addition, the copy editor, Rachelle Nashner, played a large role in improving the readability of the final version of the text. Jenny Allen went above and beyond the call of friendship in scanning all of the figures used here. The occasional technical crisis was always ably resolved at William Morris by Tali Rosenblat (present at the beginning) and Liza Gennatiempo (there at the end) and by Patrick Mulligan at Gotham Books.

My wife, Therese, has been a constant source of encouragement. She has, from its inception, been more supportive of this project than I could ever have hoped. She has read through all of the many drafts of the manuscript, and I cannot imagine this book without her editorial advice and general counsel. I am a lucky man.

INDEX

Figure 28: Photograph by Francis Simon, courtesy of AIP Emilo Segre Visual Archives, Francis Simon Collection.